Synthesis of
Electrical Networks

Synthesis of Electrical Networks

H. BAHER

Department of Electronic Engineering
University College, Dublin

A Wiley–Interscience Publication

JOHN WILEY & SONS

Chichester · New York · Brisbane · Toronto · Singapore

Library of Congress Cataloging in Publication Data:
Baher, H
    Synthesis of electrical networks.
    'A Wiley–Interscience publication.'
    Includes index.
    1. Electric network synthesis.   I. Title.
TK454.2.B33   1984   621.3815'3   83-21905

ISBN 0 471 90399 X

British Library Cataloguing in Publication Data:
Baher, H
    Synthesis of electrical networks
    1. Electric circuits
    I. Title
    621.319'2   TK454

ISBN 0 471 90399 X

Filmset and printed in Northern Ireland by The Universities Press (Belfast) Ltd.,
and bound at the Pitman Press, Bath, Avon.

Contents

v

PART III

THE AUGMENTED THEORY: SYNTHESIS OF COMMENSURATE
DISTRIBUTED AND DIGITAL NETWORKS

PART IV

APPROXIMATION THEORY AND FILTER DESIGN

Preface

This is a book on electrical network synthesis with emphasis on networks which possess filtering characteristics, and those closely related to the filter design problem. The exposition presented in the book is also of direct relevance to the synthesis of impedance transformers, broad-band matching networks, and many commonly used types of coupling networks. In fact, by virtue of common usage, the term 'filter' has become a generic name denoting a network which provides a response differing from the excitation in a prescribed manner.

At present, there are several categories of electrical networks in use, which may be classified according to the type of building blocks they employ. Examples are: conventional passive lumped networks, microwave distributed-parameter networks, digital networks, active RC networks, and analogue sampled-data networks. There are textbooks which specialize in one or other of these different categories, while occasionally devoting a brief section to a cursory discussion of the rest. Other textbooks contain material covering a number of categories, but without any unifying principle; in fact, most of these are edited volumes with unconnected contributions by different authors. The result of both approaches is that each category of network is treated as though it were, more or less, a separate discipline. Moreover, textbooks which deal with realizability theory and synthesis techniques of passive lumped networks contain material which is by now of historical interest only. These texts also lack a thorough discussion of the approximation problem in filter design. On the other hand, the excellent text *Theory of Electrical Filters* by J. D. Rhodes treats approximation theory for a wide variety of filters in a remarkably compact and unifying manner, but it is a research text which requires a detailed background in realizability theory and synthesis techniques.

This book is an attempt to put together in one volume, in a coherent manner, the synthesis techniques and approximation theory for a wide variety of networks which are commonly used in filter design practice. These include passive lumped networks, certain classes of microwave distributed networks, digital filters including wave digital structures, and a class of switched-capacitor ladder filters.

The adopted approach stems from the belief that network synthesis, rightly viewed, is a set of basic concepts and principles which are sufficiently general and well developed so as to constitute a logically self-contained discipline, regardless of the type of building blocks. In formulating the necessary theory, it is generally a defect to employ more premises than the conclusions demand. One should only employ the essential principles in virtue of which the understanding of the discipline is possible. Every contrived 'new' axiom diminishes the generality of the results while, in any discipline, the greatest possible generality is to be sought. It will be shown in this book that a considerable degree of generality is, indeed, possible if we rely heavily on the well established and rigorous theory of passive network synthesis. This theory is by now classical, but its inherent concepts and philosophy are capable of being augmented and extended to encompass a very useful and wide variety of networks, including active ones. The flexibility, consistency, and logical cohesion of the essential features of this theory make it a natural foundation for many categories of networks, and give it an almost inexhaustible potential for extension and application to other categories in the future.

So this book presents a coherent discussion of electrical network synthesis with the philosophy of passive network synthesis as a unifying theme. Emphasis is on filters and closely related topics. Realizability theory, synthesis techniques, and approximation theory are all covered. However, the treatment of approximation theory is necessarily less detailed than the rest, due to limitations on size and the availability of the text by J. D. Rhodes.

The study in this book begins where a comprehensive course in network analysis ends. The reader is, therefore, assumed to be familiar with the techniques of conventional network analysis, as well as the basic mathematical tools of matrix algebra, the Laplace transform, and complex variables. Moreover, concepts such as complex frequency, complex power, linearity, passivity, etc. are used freely without any attempt at detailed explanation. The prerequisite level of this background is represented in many references.[1-7]

The book is divided into four parts. Part I contains an introductory chapter outlining the general methodology adopted throughout the book, explaining the differences between network analysis and network synthesis, and defining the problems and philosophy of network synthesis in general. Thus, Part I is an 'overture' which prepares the reader and sets the scene for the main development in the book.

In Part II the central theme of the book is proposed and developed in relation to realizability theory and synthesis techniques of passive lumped networks. This part contains six chapters (2-7) developing the theory in a logical step-by-step fashion. This is the main part of the book which forms the foundation for later development. In selecting the material for this part, only the results which have proved to be of considerable practical and conceptual importance have been presented.

Part III contains two chapters (8-9). In the first, the central theme of Part II is augmented to accommodate a class of microwave distributed networks employing commensurate lossless transmission lines. At this point a comprehensive theory of passive network synthesis emerges which encompasses both lumped networks and very useful classes of distributed networks. In Chapter 9 a formal analogy is established between the transfer functions of digital filters and those used to characterize commensurate distributed filters. Then the synthesis techniques of digital filters are discussed with particular attention given to wave digital structures; the objective being to obtain low-sensitivity digital filters modelled on the passive ones. In line with the general philosophy of the book, the treatment of digital filters is hardware-oriented.

Part IV is the final part dealing with approximation theory and applying the synthesis techniques of the previous parts to the specific problem of filter design. This part contains four chapters (10–13). The first three chapters (10–12) deal with lumped, commensurate distributed, and digital filters, respectively. For each category, amplitude, phase, as well as combined amplitude and phase approximations are treated. Chapter 13 gives a brief discussion of a class of switched-capacitor filters which imitate the low sensitivity properties of a passive prototype filter. It is shown that the design techniques of these filters can be easily accommodated within the framework of the earlier development without any need to introduce fundamentally new network-theoretic ideas.

The text contains a large number of solved examples and there are unsolved problems at the end of the chapters. These should be regarded as part of the text, since the solutions to some of them constitute useful results. To help refresh the reader's memory, two appendices have been added, containing brief accounts of analytic functions as well as Hermitian and quadratic forms. The list of references at the end of the book is intended to be representative rather than exhaustive.

Regarding the style and presentation, a few words are in order. Due to the wide variety of networks treated in the text, a style has been adopted that is both rigorous and compact. The emphasis is on the conceptual organization of the discipline. Many concepts and definitions have been stripped of their historical associations and connotations. Consequently, some terms are defined in a rather unconventional manner.

This book is suitable as a teaching text for senior undergraduate electrical engineering and physics students taking courses in circuit theory, network synthesis, filter design, and signal processing. It should also be particularly useful for postgraduate students and researchers working in the ever-developing areas of network synthesis and signal processing. Filter design engineers in the telecommunications industry or research should find the material in this book of help.

Finally, I would like to thank my colleague, Professor Sean Scanlan. Our valuable discussions, and a decade of joint research work have contributed greatly to the material in this book.

PART I

General Introduction

OUTLINE

The various categories of networks treated in this book are reviewed. The problems of network analysis and network synthesis are defined and distinguished. The methodology and general plan of the book are explained. A list of the symbols and notation used throughout the book is given for easy reference.

Chapter 1
Setting the Scene

1.1 PERSPECTIVE

1.1.1 Passive lumped networks

The conventional passive lumped elements namely: inductors (susceptible to mutual coupling), capacitors, resistors, and transformers, have been used for a long time in the design of networks to perform a wide variety of tasks. The general properties of these networks are now well understood and the associated design techniques have reached a high degree of development and sophistication.

In this book, we are mainly concerned with networks which possess filtering properties and those which are closely related to the filter design problem. Historically, passive lumped lossless filters were the earliest types for which comprehensive design techniques exist, relying on a powerful and rigorous theory.[8] These filters have established themselves as reference designs, against which the performance of other categories is measured and compared. This is mainly due to two reasons. First, the passive lossless filter has been shown to be capable of meeting the stringent specifications to be found in all electrical engineering applications, in particular the telecommunications area. Secondly, these filters possess low sensitivity properties with respect to variations in element values; which is a highly desirable attribute from the practical viewpoint.

1.1.2 Commensurate distributed networks

As the operating frequencies increase, reaching the microwave range, the conventional lumped elements cannot be realized and the lumped concept becomes of limited use. The need, therefore, arises for different types of circuit elements capable of energy storage and transmission at these higher frequencies. The distributed-parameter concept comes to the rescue, and a very important class of distributed networks is that containing lossless transmission lines having different characteristic impedances but their lengths are integral multiples of the same length. This basic length of transmission line is called a unit element and the lines making up the entire network are said to be *commensurate* to this basic length. Coupling between

3

the lines may or may not be allowed, resulting in different subclasses of commensurate distributed networks. For these networks, it is possible to formulate a coherent theory in a manner analogous to the theory of lumped networks with the added advantage of removing the lumped constraint. This leads to a considerable increase in the richness and variety of the design techniques. In this theory, energy dissipation can still be modelled by resistors.

1.1.3 Active RC filters

Active filters employing resistors, capacitors, and active devices (usually operational amplifiers) are in common use. These were introduced to avoid the practical disadvantages of inductors at the lower end of the frequency spectrum where they become both costly and bulky. Broadly speaking, active RC filters may be divided into two main classes. The first employs design techniques which rely heavily on those of passive lumped filters, while the second has its own specific approach. Both classes are well-treated in a vast number of textbooks,[9-13] and consequently are not discussed in this book. However, the concepts of the first class will be covered by implication, whereas the second can be easily grasped by someone who studies the material in this book. Parenthetically, similar considerations apply to other categories such as crystal and mechanical filters.[14]

1.1.4 Digital filters

More recently, digital signal processing[15-17] has become an important technique in a number of areas such as the filtering of Doppler-shifted radar returns, seismic exploration, analysis of vibrations, analysis of biomedical signals, image-processing, and sonar. The increasing popularity of digital filters is due to many factors including reliability, reproducibility, high precision, and freedom from temperature and ageing effects. The decreasing cost and increased speed of operation of digital hardware, as well as the improvements in computational algorithms and software have also contributed greatly to the establishment of digital filters as viable alternatives to analogue ones, particularly in applications such as seismic exploration, where the sizes of the required lumped elements become prohibitively large as the operating frequencies become very low. Also in applications where it is required to vary the characteristics of the filter during operation such as speech processing, analogue filters cannot be used, while it is possible to achieve the required performance using digital filters. Another advantage of digital filters is that a single filter can be used to operate in several channels simultaneously. The channels share the same arithmetic elements but each requires its own memory register.

One can expect the applications of digital filters to increase as cost decreases and the design techniques continue to become widely understood.

Microprocessor-oriented implementations are also extremely powerful and accessible tools for the design of digital filters. Very often, software implementation of digital filters is both convenient and desirable.

The early design techniques of digital filters resulted in realizations which suffered from a severe sensitivity problem, in strong contrast to the excellent low sensitivity properties of passive filters. Therefore various modifications have been introduced to overcome this problem. A fundamental development led to new design techniques which result in digital structures imitating the low sensitivity properties of passive filters. These rely very heavily on the theory of commensurate distributed filters, and are called *wave digital filters*.

1.1.5 Switched-capacitor filters

Recent developments in integrated circuit technology and some novel theoretical ideas, have raised the hopes of obtaining low sensitivity analogue filters, operating in the sampled-data mode, and capable of being accurately and conveniently realized in integrated circuit form. The latest examples of these employ operational amplifiers, capacitors, and analogue switches as the basic building blocks.[18] These have come to be known as *switched-capacitor filters* and are at present under thorough investigation by circuit theorists and technologists. In fact, until recently it was not possible to realize analogue filters in silicon integrated form. Active filters could only be realized in hybrid form using thick- and thin-film techniques. Therefore, the designer could not make use of the rapid advances in very large scale integration (VLSI) circuit technology. With the advent of switched-capacitor filters, low-power metal-oxide semiconductor (MOS) silicon-integrated filter chips are now available. Moreover, these offer the prospect of integrating entire information systems on a single chip. So far, the most promising types of switched-capacitor filters are those modelled on passive filters, thus retaining the good sensitivity properties of the latter.

1.1.6 Remarks

From the above brief survey, a number of points arise. First, there are several categories of commonly used networks which may be classified according to the type of building blocks they employ. Secondly, the well-established design techniques of passive networks play the central role in formulating the design techniques for other categories of useful filters, including active ones. These considerations dictate the general philosophy of this book, which will be discussed shortly. But first we give a brief account of passive network theory.

1.2 PASSIVE NETWORK THEORY

The essential principles of passive network synthesis are used throughout this book as a central theme which is later augmented and extended to

6

encompass other categories of networks. Therefore, we now give a brief outline of the objectives and philosophy of the two main disciplines in passive network theory, namely: network analysis and network synthesis. It is of paramount importance that the reader should appreciate the difference between the techniques and methodology of network analysis on the one hand, and network synthesis on the other.

1.2.1 Network analysis

In the problem of network analysis, a complete network is given which contains (idealized) passive elements and (idealized) voltage and current sources (Fig. 1.1(a)). It is then required to calculate the voltages and currents at various points in the network. This is accomplished by deducing from the given network a set of equations, the solution of which gives the required quantities; the reader is undoubtedly familiar with the methods of nodal analysis and loop analysis, etc. The currents and voltages throughout the network are, in general, functions of the complex frequency variable p,

$$p = \sigma + j\omega \tag{1.1}$$

It is then possible to separate the variables (i.e. voltages and currents) into two sets. The first contains the so-called port variables which are the voltages and currents at a number n of accessible terminal-pairs, each known as a port. The second set of variables contains the so-called internal variables. Now, the internal variables can be eliminated from all the equations and the remaining set of equations relates the port variables among themselves. It is also possible to select the accessible ports such that sources can exist only at the ports while the rest of the network is passive, as shown in Fig. 1.1(b). Thus a passive n-port network results, known as an n-port for short, whose external behaviour, relative to the ports, is defined once we know the set of equations relating the voltages and currents at the ports.

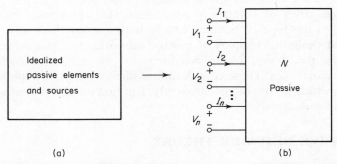

(a)   (b)

Fig. 1.1   (a) A complete network: interconnection of idealized passive elements and sources, (b) a passive n-port.

Networks can, therefore, be classified in terms of the number of accessible ports. Let

$$[V(p)] = \begin{bmatrix} V_1(p) \\ V_2(p) \\ \vdots \\ V_n(p) \end{bmatrix} \tag{1.2}$$

$$[I(p)] = \begin{bmatrix} I_1(p) \\ I_2(p) \\ \vdots \\ I_n(p) \end{bmatrix} \tag{1.3}$$

be the column matrices whose entries are the port-voltages and port-currents, respectively. Then these can be related by

$$[V(p)] = [Z(p)][I(p)] \tag{1.4}$$

where

$$[Z(p)] = \begin{bmatrix} z_{11}(p) & z_{12}(p) & \cdots & z_{1n}(p) \\ z_{21}(p) & z_{22}(p) & \cdots & z_{2n}(p) \\ \vdots & & & \vdots \\ z_{n1}(p) & \cdots & & z_{nn}(p) \end{bmatrix} \tag{1.5}$$

which is the (open-circuit) impedance matrix of the n-port. Alternatively, we may write

$$[I(p)] = [Y(p)][V(p)] \tag{1.6}$$

where

$$[Y(p)] = \begin{bmatrix} y_{11}(p) & y_{12}(p) & \cdots & y_{1n}(p) \\ y_{21}(p) & y_{22}(p) & \cdots & y_{2n}(p) \\ \vdots & & & \vdots \\ y_{n1}(p) & \cdots & & y_{nn}(p) \end{bmatrix} \tag{1.7}$$

which is the (short-circuit) admittance matrix of the n-port. Clearly from (1.4) and (1.6),

$$[Z(p)] = [Y(p)]^{-1}, \qquad [Y(p) = [Z(p)]^{-1} \tag{1.8}$$

It is worth noting that the existence of $[Z(p)]$ or $[Y(p)]$ is not guaranteed for any network.[19,20] Indeed, there are networks which possess neither a $[Z(p)]$ nor a $[Y(p)]$ representation.

Two special cases are of major importance to the material in this book.

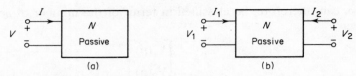

Fig. 1.2   The special cases of major interest. (a) A passive one-port, (b)
a passive two-port.

These are the one-port and two-port networks shown in Fig. 1.2. In the first
case the voltage and current variables are related by a scalar (function),

$$V(p) = Z(p)I(p) \tag{1.9}$$

or

$$I(p) = Y(p)V(p) \tag{1.10}$$

where $Z(p)$ and $Y(p)$ are the driving-point impedance and admittance of the
one-port, respectively. For the two-port shown in Fig. 1.2(b), there are
several ways of relating the port variables, with some of which the reader is
undoubtedly familiar. Let $n = 2$ in (1.2) to (1.7) so that

$$[V(p)] = \begin{bmatrix} V_1(p) \\ V_2(p) \end{bmatrix} \tag{1.11}$$

and

$$[I(p)] = \begin{bmatrix} I_1(p) \\ I_2(p) \end{bmatrix} \tag{1.12}$$

We may then write (1.4) to (1.5) with $n = 2$ as

$$\begin{bmatrix} V_1(p) \\ V_2(p) \end{bmatrix} = \begin{bmatrix} z_{11}(p) & z_{12}(p) \\ z_{21}(p) & z_{22}(p) \end{bmatrix} \begin{bmatrix} I_1(p) \\ I_2(p) \end{bmatrix} \tag{1.13}$$

Therefore, the impedance matrix of the two-port,

$$[Z(p)] = \begin{bmatrix} z_{11}(p) & z_{12}(p) \\ z_{21}(p) & z_{22}(p) \end{bmatrix} \tag{1.14}$$

is a 2×2 matrix which characterizes the external behaviour of the two-port.
The entries of $[Z(p)]$ are the z-parameters, which from (1.13) are defined as

$$z_{11}(p) = \frac{V_1(p)}{I_1(p)} \bigg|_{I_2=0}$$

$$z_{22}(p) = \frac{V_2(p)}{I_2(p)} \bigg|_{I_1=0}$$

$$z_{12}(p) = \frac{V_1(p)}{I_2(p)} \bigg|_{I_1=0} \tag{1.15}$$

$$z_{21}(p) = \frac{V_2(p)}{I_1(p)} \bigg|_{I_2=0}$$

Hence $z_{11}(p)$ and $z_{22}(p)$ are *driving-point* impedances, since each function is the ratio of a voltage to a current at the same port. On the other hand, $z_{12}(p)$ and $z_{21}(p)$ are *transfer* impedances, since each function is the ratio of a voltage to a current at different ports.

Similarly the admittance matrix of a two-port relates the port variables as

$$\begin{bmatrix} I_1(p) \\ I_2(p) \end{bmatrix} = \begin{bmatrix} y_{11}(p) & y_{12}(p) \\ y_{21}(p) & y_{22}(p) \end{bmatrix} \begin{bmatrix} V_1(p) \\ V_2(p) \end{bmatrix} \tag{1.16}$$

Therefore, the admittance matrix

$$[Y(p)] = \begin{bmatrix} y_{11}(p) & y_{12}(p) \\ y_{21}(p) & y_{22}(p) \end{bmatrix} \tag{1.17}$$

is a 2×2 matrix which characterizes the external behaviour of the two-port. The entries of $Y(p)$ are the y-parameters defined by (1.16) as

$$y_{11}(p) = \frac{I_1(p)}{V_1(p)} \bigg|_{V_2=0}$$

$$y_{22}(p) = \frac{I_2(p)}{V_2(p)} \bigg|_{V_1=0}$$

$$y_{12}(p) = \frac{I_1(p)}{V_2(p)} \bigg|_{V_1=0} \tag{1.18}$$

$$y_{21}(p) = \frac{I_2(p)}{V_1(p)} \bigg|_{V_2=0}$$

Hence $y_{11}(p)$ and $y_{22}(p)$ are *driving-point* admittances while $y_{12}(p)$ and $y_{21}(p)$ are *transfer* admittances.

An alternative way to relate the port variables for the two-port in Fig. 1.2(b) is to write

$$\begin{bmatrix} V_1(p) \\ I_1(p) \end{bmatrix} = \begin{bmatrix} A(p) & B(p) \\ C(p) & D(p) \end{bmatrix} \begin{bmatrix} V_2(p) \\ -I_2(p) \end{bmatrix} \tag{1.19}$$

where the matrix

$$[T(p)] = \begin{bmatrix} A(p) & B(p) \\ C(p) & D(p) \end{bmatrix} \tag{1.20}$$

is called the *transmission* matrix (or *chain* matrix) of the two-port. Its entries are defined by (1.19) as

$$A(p) = \frac{V_1(p)}{V_2(p)} \bigg|_{I_2=0}$$

$$B(p) = \frac{V_1(p)}{-I_2(p)} \bigg|_{V_2=0}$$

$$C(p) = \frac{I_1(p)}{V_2(p)} \bigg|_{I_2=0} \tag{1.21}$$

$$D(p) = \frac{I_1(p)}{-I_2(p)} \bigg|_{V_2=0}$$

which are all *transfer* functions. The description of the two-port in terms of its transmission matrix is convenient when several networks are connected in cascade. The overall transmission matrix is then equal to the product of the transmission matrices of the individual two-ports, in the same order. It is also useful for the description of the ideal transformer which does not possess either a [Z] or [Y] representation.

So, *in the context of the external behaviour* of a network, the problem of analysis may now be defined accurately, for the two cases of major interest.

(a) For a passive one-port as shown in Fig. 1.2(a), the problem of network analysis consists in finding the response (i.e. $I(p)$ or $V(p)$) of the given network due to the given excitation $(V(p)$ or $I(p))$. Thus the problem is solved once we have obtained the driving-point impedance or admittance of the one-port, since the excitation and response are related by (1.9) or (1.10).

(b) For a passive two-port as shown in Fig. 1.2(b), network analysis consists in finding the response (currents or voltages at the ports) due to the given excitation (voltages or currents at the ports) of the given network. Thus, the problem is solved once we have obtained a characterization of the two-port in terms of one of its defining matrices, e.g. [Z], [Y], or [T]. Any of these relate the excitation to the response by (1.13), (1.16), or (1.19).

Similar statements apply to the most general passive n-port of Fig. 1.1(b).[19-21] But throughout this book we are mainly interested in one-ports and two-ports. Hence our discussion here, will be largely confined to these cases.

Now, the central problem in passive filter design is concerned with a lossless two-port operating between a resistive source and a resistive load as shown in Fig. 1.3. Later in the book, we shall see that the most convenient and meaningful description of this doubly terminated lossless two-port, is in terms of the *scattering matrix*. This will be discussed in detail, but for the moment let it only be mentioned that the transfer function of major interest is given by

$$S_{21}(p) = 2\sqrt{\frac{r_g}{r_\ell}} \frac{V_2(p)}{V_g(p)} \tag{1.22}$$

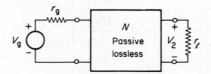

Fig. 1.3 The doubly terminated loss-less two-port.

and in the sinusoidal steady state $(p = j\omega)$,

$$|S_{21}(j\omega)|^2 = \frac{|V_2(j\omega)|^2/r_\ell}{|V_g(j\omega)|^2/4r_g} \qquad (1.23)$$

which is the ratio of the power delivered to the load to the maximum power available from the source. Therefore, expression (1.23) determines the amplitude response of the network. We may also write

$$S_{21}(j\omega) = |S_{21}(j\omega)| \, e^{j\psi(\omega)} \qquad (1.24)$$

where $\psi(\omega)$ is the phase function. If the network is given, then we can find its amplitude and phase characteristics, and this problem belongs to the discipline of network analysis.

1.2.2 Network synthesis

We now turn to the problem of network synthesis, the subject of this book. This may be defined as the problem of constructing a network having a prescribed external behaviour relative to a number of ports. Let us be more specific and consider network synthesis for the cases of major interest.

(a) In the case of a one-port, given the driving-point impedance $Z(p)$ (or admittance $Y(p)$) which characterizes the one-port, we are required to find a network whose external behaviour is described by the given $Z(p)$.

(b) In the case of a two-port, given one of the matrices which characterize the two-port (e.g. $[Z(p)]$, $[Y(p)]$, or $[T(p)]$) we are required to find the network whose external behaviour, at the ports, may be described by the given matrix.

When the synthesis problem is posed in the way outlined above, various points automatically arise.

(i) The problem may not have a solution in terms of the physically realizable passive components. This becomes evident if we consider the trivial example

$$Z(p) = p - \frac{1}{p} - 2$$

which requires a negative capacitor and a negative resistor. When the expression for the given function, or matrix, becomes more elaborate, one cannot rely on inspection to test for physical realizability. Hence, before attempting to find the network from the given characterization, we must have at our disposal some rigorous means for finding out whether the given function, or matrix, is realizable as the interconnection of physical components of the specified type. This preliminary step is the subject of *realizability theory*.

(ii) The specified function, or matrix, does not define a single network, but rather *a class of equivalent networks* which may be described by the same function, or matrix. Thus, the solution to the synthesis problem, if it exists, is *not unique* (by contrast with the analysis problem). The different networks which have the same external behaviour may have different topologies as well as different element values.

(iii) The synthesis problem is essentially a design problem, i.e. the construction of a network to perform a certain task. Therefore the function or matrix to be realized is not given in practice, but must also be *determined* before the synthesis is to be performed, in such a way that the resulting network performs the required task. In particular, our interest is concentrated in the design of the doubly terminated lossless two-port of Fig. 1.3, such that its transfer characteristics meet certain specifications. These specifications must first be expressed in mathematical form by deriving a specific transfer function as defined in (1.22). Then, a characterization of the two-port in terms of one of the describing matrices (e.g. $[Z(p)]$, $[Y(p)]$, or $[T(p)]$) is obtained. This must be obtained in such a way as to guarantee physical realizability. In other words, the characterizing function or matrix must, on the one hand meet the design specifications subject to a certain optimality criterion, and on the other it must also satisfy the necessary and sufficient conditions for realizability.

Now although the study in this book includes active networks, the adopted approach to the entire synthesis problem takes as its foundations the techniques of passive network synthesis. The following three closely related stages define the procedure of network synthesis together with its subsidiary problems.

(a) The derivation of the set of necessary and sufficient conditions under which a given function, or matrix, is realizable by a network using the specified types of building blocks, and possibly in a particular structure. For example, in the case of passive lumped networks the building blocks are the conventional circuit elements.

(b) The selection of a subclass of functions or matrices, belonging to the realizable class defined in (a), which also satisfies certain requirements regarding its performance, e.g. amplitude and/or phase characteristics.

(c) The realization of the network in the preferred structure, i.e. finding the structure and element values of the network whose characterization was obtained in (b).

Although the term *synthesis* may be justly applied to the above three interwoven stages, for the sake of clarity the problems defined in (a) and (c) will be referred to as the *synthesis*, whereas step (b) will be termed the *approximation problem*. These will be treated separately in this book, but with the proviso that when one eye is on the synthesis the other must be

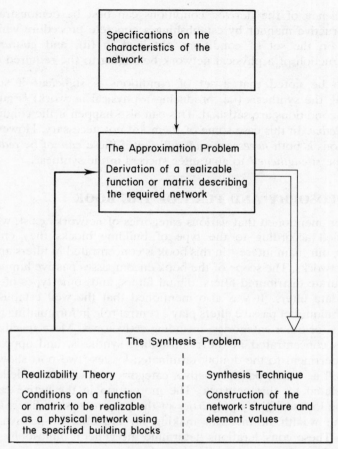

Fig. 1.4 The divisions and subsidiary problems of network synthesis.

kept on the approximation. The outline and divisions of the procedure are depicted in Fig. 1.4.

Now, in order to derive the necessary and sufficient conditions for a given function, or matrix, to represent a physical network, the following steps constitute a logical and well established procedure.

(i) The basic building blocks are selected and regarded as admissible.
(ii) Network analysis is used to obtain the general mathematical properties of the arbitrary interconnection of such building blocks. At this stage, we can assert that a network exists *only if* the given function, or matrix, satisfies the derived set of mathematical conditions, i.e. these conditions are *necessary* for realizability. But we are by no means justified in assuming the *sufficiency* of the conditions which have been derived by analysis.

(iii) *Sufficiency* of the derived conditions can best be demonstrated in a constructive manner by establishing a concrete procedure which relies only on the set of conditions obtained in (ii), and *guarantees* the construction of a physical network belonging to the required class.

It is to be noted that a set of conditions is *sufficient* if success in performing the synthesis (i.e. producing a physical network) is guaranteed when the conditions are satisfied. This can also happen if the conditions are *over-restrictive*. In this case some of them are not necessary. However, a set of conditions is both *necessary and sufficient* if these *cannot* be *reduced* and *need not* be *strengthened* to guarantee success in the synthesis.

1.3 PHILOSOPHY AND PLAN OF THE BOOK

It has been mentioned that various categories of networks exist, which may be classified according to the type of building blocks they employ. In particular, our main interest in this book is concentrated in filters and closely related networks. The scope of the book encompasses passive lumped filters, commensurate distributed filters, digital filters, and some types of analogue sampled-data filters. It was also mentioned that the well established synthesis techniques of passive filters play a central role in formulating the design techniques for other categories, including active ones. More specifically, our interest is concentrated in the realizability, synthesis, and approximation methods pertinent to the doubly terminated lossless two-port shown in Fig. 1.3, as well as other, possibly active, categories which are modelled on, or closely related to, this network. The motivation in the active cases is to imitate the low sensitivity properties of the passive lossless filter, as well as to use the wealth of material making up the area of passive network synthesis. These considerations determine the general philosophy and plan of this book, and are now outlined.

Instead of regarding each category of networks as a separate discipline with its own design principles and hence to be discussed independently, a different approach is adopted. A common set of principles, for the synthesis of all defined categories of networks, is sought. This is proposed as a central theme which puts together the essential features of network synthesis in a coherent manner. This objective may, in principle, be contemplated in two different ways.

 (i) A general *abstract* theory is formulated, which is independent of the type of building blocks being used. Then, the theory is applied to the synthesis of the different categories of networks separately. This approach may be called *deductive*.
(ii) Alternatively, the proposed theory may first be developed in relation to a specific category of networks. Then, a formal analogy is established between this specific category and other defined categories. Finally it is shown how the same theory may be augmented, modified, and extended

to encompass the networks which employ different types of building blocks. So the generality of the essential features of this theory increases with every new category that it encompasses. In its broad outlines, this approach may be called *inductive*.

In this book, the inductive approach outlined in (ii) above is adopted. The central theme is developed in relation to passive lumped networks, then later augmented and extended to other categories namely: commensurate distributed, digital, and some types of switched-capacitor networks. The central theme emphasizes the doubly terminated lossless two-port of Fig. 1.3; naturally the singly terminated and unterminated lossless two-port cases are covered by implication. The general plan of the book is depicted in Fig. 1.5.

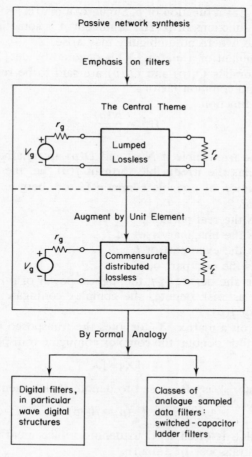

Fig. 1.5 General plan and philosophy of the book.

The adopted approach has several advantages over the alternative (in fact, only *speculative*) abstract deductive method. First, at every stage of the inductive development we deal with a specific concrete set of building blocks which are defined idealized physical components. Secondly, the adopted approach allows greater flexibility in the reading and teaching of the material. Finally, it is more suitable for engineering students. Throughout the book, emphasis is on the conceptual organization of the whole discipline, and care is taken not to interrupt the continuity of the development.

1.4 NOTATION

The notation used in this book is largely standard, and every symbol is explained as it is first introduced in the text. However, for easy reference, below is a list of the symbols and notation used throughout.

(i) $f(p)$ means f as a function of p. Similarly a matrix $[X(p)]$ is one whose entries are functions of p. The argument p is sometimes dropped for convenience, where no ambiguity may arise.

(ii) The determinant of a matrix $[X]$ is denoted by $\det[X]$.

(iii) Two polynomials $Q_1(p)$ and $Q_2(p)$ are said to be relatively prime, if they have no common factor.

(iv) A rational function

$$f(p) = \frac{N(p)}{D(p)}$$

is said to be irreducible if $N(p)$ and $D(p)$ are relatively prime.

(v) Ir $f(p)$ means the irreducible form of $f(p)$, i.e. the expression after the cancellation of possible common factors between numerator and denominator.

(vi) Re f means the real part of f.
Im f means the imaginary part of f.
Ev f means the even part of f.
Od f means the odd part of f.
deg f means the degree of f, the highest power of p in f.

(vii) The upper asterisk denotes the complex conjugate, $f^*(p) \equiv$ complex conjugate of $f(p)$.

(viii) The prime on a matrix $[X]'$ denotes the transposed matrix.

(ix) The upper tilde denotes the complex conjugate transpose of a matrix,

$$[\tilde{X}] \equiv [X^*]'$$

(x) A lower asterisk is often used to denote replacement of p by $-p$

$$f_*(p) \equiv f(-p)$$

(xi) A lower tilde is often used to denote replacement of p by $-p$ and taking the transpose of a matrix,

$$[\underset{\sim}{X}(p)] \equiv [X_*(p)]' \equiv [X(-p)]'.$$

PART II

The Central Theme: Synthesis of Passive Lumped Networks

OUTLINE

The realizability theory and synthesis techniques of passive lumped linear and time-invariant networks, are developed. This part contains six chapters formulating the theory in a logical step-by-step fashion. The main interest is concentrated in one-ports and two-ports due to their direct applicability to the design of filters and closely related networks. This is the main part of the book, providing the foundation of later developments.

Chapter 2
The Key Concepts: Fundamental Properties of Passive One-ports

2.1 INTRODUCTION

This chapter is concerned with the realizability conditions of driving-point functions which characterize passive, lumped, linear, and time-invariant networks. First, the basic building blocks are selected and defined. Then the chapter proceeds by gradually introducing the key concepts in relation to the realizability of driving-point impedances of passive one-ports. The necessity of a mathematical property known as the *positive real* condition is shown to follow from energy considerations and the passivity requirement. Some related concepts, in particular *Hurwitz polynomials* and the *bounded real* condition, are discussed. Next the description of a one-port in terms of its input reflection coefficient is presented, thus introducing the *scattering description* of networks. Finally a complete procedure is given for testing a function for positive real character. The chapter concludes with a summary of the salient results.

2.2 THE BASIC BUILDING BLOCKS

The networks discussed in this part of the book are mainly those containing the conventional circuit elements, namely, resistors, capacitors, inductors (susceptible to mutual coupling), and ideal transformers. These are shown in Fig. 2.1. In terms of the complex frequency variable p, the voltage-current relations of these elements are as follows,

(i) For a resistor

$$v(p) = Ri(p) \tag{2.1}$$

(ii) For a capacitor

$$v(p) = \frac{1}{Cp} i(p) \tag{2.2}$$

19

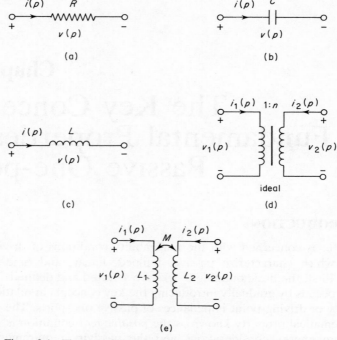

Fig. 2.1 The conventional circuit elements. (a) Resistor,
(b) capacitor, (c) inductor, (d) ideal transformer, (e) coupled coils.

(iii) For an inductor

$$v(p) = Lpi(p) \tag{2.3}$$

(iv) For an ideal $1:n$ transformer

$$\begin{bmatrix} v_1(p) \\ i_1(p) \end{bmatrix} = \begin{bmatrix} \dfrac{1}{n} & 0 \\ 0 & n \end{bmatrix} \begin{bmatrix} v_2(p) \\ -i_2(p) \end{bmatrix} \tag{2.4}$$

(v) For a pair of mutually coupled coils as shown in Fig. 2.1(e), we may write

$$\begin{bmatrix} v_1(p) \\ v_2(p) \end{bmatrix} = p \begin{bmatrix} L_1 & M \\ M & L_2 \end{bmatrix} \begin{bmatrix} i_1(p) \\ i_2(p) \end{bmatrix} \tag{2.5}$$

and from energy considerations the inductance matrix must be positive semi-definite (see Appendix A.1). Thus for the non-degenerate case in which $L_1 \neq 0$, $L_2 \neq 0$ we must have

$$L_1 > 0 \tag{2.6a}$$

$$L_1 L_2 - M^2 \geq 0 \tag{2.6b}$$

and $L_2 > 0$ is a consequence of the above conditions. If (2.6b) is satisfied with equality, the coils are said to be *perfectly coupled* and we have

$$M = \pm\sqrt{L_1 L_2} \qquad (2.7)$$

Letting

$$n = \pm\sqrt{\frac{L_2}{L_1}} = \frac{M}{L_1} = \frac{L_2}{M} \qquad (2.8)$$

we have

$$M = nL_1 \qquad (2.9)$$

Now, the transmission matrix of the inductance two-port with perfect coupling as shown in Fig. 2.2(a) is given by

$$[T] = \begin{bmatrix} \dfrac{L_1}{M} & 0 \\ \dfrac{1}{pM} & \dfrac{M}{L_1} \end{bmatrix} \qquad (2.10a)$$

and using (2.9) we have

$$[T] = \begin{bmatrix} \dfrac{1}{n} & 0 \\ \dfrac{1}{nL_1 p} & n \end{bmatrix} \qquad (2.10b)$$

Next consider the cascade connection of a shunt inductor L_1 and an ideal $1:n$ transformer, as shown in Fig. 2.2(b). The transmission matrix of the cascade is given by

$$[T_1] = \begin{bmatrix} 1 & 0 \\ \dfrac{1}{pL_1} & 1 \end{bmatrix} \begin{bmatrix} \dfrac{1}{n} & 0 \\ 0 & n \end{bmatrix}$$

$$= \begin{bmatrix} \dfrac{1}{n} & 0 \\ \dfrac{1}{nL_1 p} & n \end{bmatrix} \qquad (2.11)$$

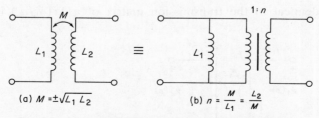

(a) $M = \pm\sqrt{L_1 L_2}$ (b) $n = \dfrac{M}{L_1} = \dfrac{L_2}{M}$

Fig. 2.2   (a) Perfectly-coupled coils, (b) equivalent circuit.

But the above matrix is identical to that in (2.10). Therefore, a pair of perfectly coupled coils are equivalent to a self-inductance and an ideal transformer. Thus, the theory of networks composed of resistors, capacitors, (self) inductors, and perfectly-coupled coils, can be reduced to one of networks containing resistors, capacitors, inductors (with self-inductance only), and ideal transformers.

We are mainly interested in reciprocal networks containing only the elements just described. However, for the sake of completeness, and to avoid any loss of generality, we introduce the ideal gyrator: an extra non-reciprocal element first postulated by Tellegen.[22] We shall have occasion later on in the book to add it to the basic building blocks making up the network. The ideal gyrator is the two-port element shown in Fig. 2.3, and is defined by the equations,

$$v_1(p) = -\alpha i_2(p)$$
$$v_2(p) = \alpha i_1(p)$$

(2.12)

where α is the gyration resistance. The behaviour defined by (2.12) can be approximated by certain electromechanical or microwave devices. The impedance and transmission matrices of an ideal gyrator are given by

$$[Z] = \begin{bmatrix} 0 & -\alpha \\ \alpha & 0 \end{bmatrix}$$

(2.13)

$$[T] = \begin{bmatrix} 0 & \alpha \\ \dfrac{1}{\alpha} & 0 \end{bmatrix}$$

(2.14)

from which the element is clearly non-reciprocal since $z_{12} \neq z_{21}$. A further interesting property of such an element can be seen by considering the cascade connection of two ideal gyrators of gyration resistance α_a and α_b as shown in Fig. 2.4(a). The transmission matrices of the two gyrators are multiplied to obtain the transmission matrix of the cascade. Thus

$$[T] = \begin{bmatrix} \dfrac{\alpha_a}{\alpha_b} & 0 \\ 0 & \dfrac{\alpha_b}{\alpha_a} \end{bmatrix}$$

(2.15)

which is identical to the transmission matrix of a $1:(\alpha_b/\alpha_a)$ ideal trans-

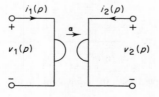

Fig. 2.3   Ideal gyrator.

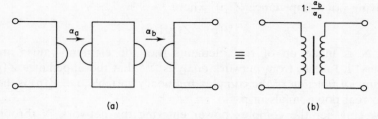

Fig. 2.4   (a) Two cascaded gyrators, (b) equivalent circuit.

former. Thus, two cascaded ideal gyrators are equivalent to an ideal trans-former, as shown in Fig. 2.4(b).

2.3   ENERGY CONSIDERATIONS

As explained in Chapter 1, a one-port network (or 'one-port' for short) is a network to which we have access by means of one pair of terminals known as the port. Thus, the external behaviour of the network relative to the port is describable by a single function of frequency: its driving-point impedance $Z(p)$ or admittance $Y(p)$. We now take our first steps towards obtaining an answer to the fundamental question:

What are the necessary and sufficient conditions under which a given function $Z(p)$ is realizable as the driving-point impedance of a passive, lumped, linear, and time-invariant network?

To this end, network analysis is used to establish the *necessity* of a certain mathematical condition for the realizability of a given impedance (or admittance) function. However, for a proof of the *sufficiency* of this realizability condition the reader must wait until a much later chapter. For such a proof requires the demonstration of a complete synthesis technique of the most general type of impedance, satisfying only the fundamental realizability condition.

Consider the one-port N shown in Fig. 2.5, which is assumed passive, and a source can exist only at the port. The given network may be described by

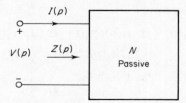

Fig. 2.5   A passive one-port.

its driving-point impedance $Z(p)$, where

$$V(p) = Z(p)I(p) \qquad (2.16)$$

Since N is made up of real elements (i.e. the element values are real numbers) it follows from network analysis[20,21] that the impedance $Z(p)$ is a *real rational function* of the complex frequency variable, i.e. $Z(p)$ is the ratio of two real polynomials in p.

Now consider the complex power entering the network N through the port. This is given by

$$W(p) = V(p)I^*(p) \qquad (2.17)$$

where the upper asterisk denotes the complex conjugate. Letting $v(p)$ and $i(p)$ be the voltage and current associated with a typical resistor, capacitor, or inductor of the passive one-port, then the complex power in such an element is

$$w(p) = v(p)i^*(p) \qquad (2.18)$$

For a resistor we have for the average dissipated power,

$$w_R(p) = Ri(p)i^*(p) \qquad (2.19)$$

For a capacitor C, the average electric energy is

$$\varepsilon_C(p) = \tfrac{1}{2}Cv(p)v^*(p) \qquad (2.20)$$

Combining (2.2) with (2.18) and using (2.20) we obtain for the complex power entering a capacitor,

$$
\begin{aligned}
w_c(p) &= Cp^*v(p)v^*(p) \\
&= 2p^*\varepsilon_c(p)
\end{aligned}
\qquad (2.21)
$$

In an inductor L, the average magnetic energy is given by

$$\varepsilon_L(p) = \tfrac{1}{2}Li(p)i^*(p) \qquad (2.22)$$

Combining (2.3) with (2.18) and using (2.22) we obtain for the complex power entering an inductor,

$$
\begin{aligned}
w_L(p) &= Lpi(p)i^*(p) \\
&= 2p\varepsilon_L(p)
\end{aligned}
\qquad (2.23)
$$

For a pair of mutually coupled coils as shown in Fig. 2.1(e) the average magnetic energy is given by

$$
\begin{aligned}
\varepsilon_{L,M}(p) &= \tfrac{1}{2}L_1 i_1(p)i_1^*(p) + \tfrac{1}{2}L_2 i_2(p)i_2^*(p) \\
&\quad + \tfrac{1}{2}M\{i_1(p)i_2^*(p) + i_1^*(p)i_2(p)\} \\
&= \varepsilon_{L_1} + \varepsilon_{L_2} + \varepsilon_M
\end{aligned}
\qquad (2.24)
$$

or

$$\varepsilon_{L,M}(p) = \tfrac{1}{2}[i_1^*(p) \quad i_2^*(p)]\begin{bmatrix} L_1 & M \\ M & L_2 \end{bmatrix}\begin{bmatrix} i_1(p) \\ i_2(p) \end{bmatrix} \qquad (2.25)$$

In the above expression, passivity requires that the total magnetic energy $\varepsilon_{L,M}(p)$ must be non-negative for all possible values (not both zero) of $i_1(p)$ and $i_2(p)$. Thus the quadratic form on the right-hand side of (2.25) must be positive semi-definite (see Appendix A.1). This means that the inductance matrix, in the non-degenerate case with $L_1 \neq 0$ and $L_2 \neq 0$, must satisfy

$$L_1 > 0 \tag{2.26a}$$

$$L_1 L_2 - M^2 \geq 0 \tag{2.26b}$$

from which it also follows that $L_2 > 0$. The coils are perfectly coupled if (2.26b) is satisfied with the equality. Combining (2.5) with (2.24) we obtain for the complex power entering the pair of coils,

$$w_{L,M} = 2p\varepsilon_{L,M}(p) \tag{2.27}$$

It must also be clear that the ideal transformer of Fig. 2.1(d) absorbs no power, since the total complex power entering the element through both ports is given by

$$w_T(p) = i_1^*(p)v_1(p) + i_2^*(p)v_2(p)$$

which upon use of (2.4) gives

$$w_T(p) = 0 \tag{2.28}$$

Now let the sum of the average dissipated power for all elements in the network be $W_d(p)$, the sum of the average electric energy for all elements be $T_e(p)$, and the sum of the average magnetic energy for all elements be $T_m(p)$, i.e.

$$W_d(p) = \Sigma w_R(p) \tag{2.29}$$

$$T_e(p) = \Sigma \varepsilon_c(p) \tag{2.30}$$

$$T_m(p) = \Sigma \varepsilon_L(p) + \Sigma \varepsilon_{L,M}(p) \tag{2.31}$$

A classical result[20,23] allows us to assert that *the passive network N together with the source conserve complex power*. This means that the total complex power entering N through the port, is equal to the sum of all complex powers in the elements for all values of the complex frequency variable p. Using this result we have from (2.17), (2.19), (2.20), (2.23), (2.25), (2.27), and (2.29) to (2.31)

$$W(p) = V(p)I^*(p)$$
$$= W_d(p) + 2pT_m(p) + 2p^*T_e(p) \tag{2.32}$$

Examination of expressions (2.19) to (2.27) shows that $W_d(p)$, $T_m(p)$, and $T_e(p)$ are all real and positive quantities for all p. Therefore (note that $p = \sigma + j\omega$) we may write for the total complex power,

$$W(p) = W_d(p) + 2\sigma\{T_m(p) + T_e(p)\}$$
$$+ 2j\omega\{T_m(p) - T_e(p)\} \tag{2.33}$$

whose real part is given by

$$\text{Re } W(p) = W_d(p) + 2\sigma\{T_m(p) + T_e(p)\} \tag{2.34}$$

The *crucial property* of the above expression is that it is *positive for values of p with positive real part*, i.e. for $\sigma > 0$. Thus,

$$\text{Re } W(p) > 0 \qquad \text{for Re } p > 0 \tag{2.35}$$

As a short digression, let us allow ideal gyrators to exist in the network. The complex power absorbed by the ideal gyrator of Fig. 2.3 through both ports is given by

$$w_g(p) = i_1^*(p)v_1(p) + i_2^*(p)v_2(p)$$

which upon use of (2.12) becomes,

$$w_g(p) = \alpha\{i_2^*(p)i_1(p) - i_1^*(p)i_2(p)\} \tag{2.36}$$

which is purely imaginary. Hence, the ideal gyrator absorbs no real power and is lossless. Therefore, if we allow ideal gyrators to exist in addition to the conventional elements, we conclude that the real part of the complex power as given by (2.34) remains unchanged and still satisfies (2.35).

It is our objective at this stage to translate this property of Re $W(p)$ as expressed in (2.35) into a corresponding condition on the driving-point impedance of the one-port. Substituting from (2.16) into (2.32) we have

$$Z(p)I(p)I^*(p) = W_d(p) + 2\sigma\{T_m(p) + T_e(p)\}$$
$$+ 2j\omega\{T_m(p) - T_e(p)\} \tag{2.37}$$

Taking the conjugate of both sides of (2.37), and noting that $W_d(p)$, $T_m(p)$, and $T_e(p)$ are all real quantities for all p, we have

$$Z^*(p)I(p)I^*(p) = W_d(p) + 2\sigma\{T_m(p) + T_e(p)\}$$
$$- 2j\omega\{T_m(p) - T_e(p)\} \tag{2.38}$$

Addition of (2.37) and (2.38) gives

$$\tfrac{1}{2}\{Z(p) + Z^*(p)\}I(p)I^*(p) = W_d(p) + 2\sigma\{T_m(p) + T_e(p)\} \tag{2.39}$$

But we have seen that the right-hand side of (2.39) satisfies (2.35), i.e. it is positive for values of p with positive real part. Therefore the left-hand side of (2.39) has the same property for all possible non-zero values of $I(p)$. Noting that $I(p)I^*(p) = |I(p)|^2$, which is real and positive for all p, then we must have

$$\tfrac{1}{2}\{Z(p) + Z^*(p)\} > 0 \qquad \text{for Re } p > 0 \tag{2.40}$$

or

$$\text{Re } Z(p) > 0 \qquad \text{for Re } p > 0 \tag{2.41}$$

Definition 2.1 A real function whose real part is positive for values of p with positive real part is called a *positive real function*, abbreviated p.r.f.

Using similar analysis, it may be shown that the driving-point admittance of the one-port satisfies

$$\tfrac{1}{2}\{Y(p) + Y^*(p)\}V(p)V^*(p) = W_d(p) + 2\sigma\{T_m(p) + T_e(p)\} \qquad (2.42)$$

from which the admittance $Y(p)$ is also a positive real function.

2.4 THE POSITIVE REAL CONDITION

The results of our development so far may now be summarized.

Theorem 2.1 *The driving-point impedance $Z(p)$ of a passive, lumped, linear, and time-invariant network, is a positive real function (p.r.f.) i.e.,*

(i) $\qquad\qquad\qquad\qquad Z(p)$ *is real for p* real $\qquad\qquad\qquad\qquad (2.43)$

(ii) $\qquad\qquad\qquad\qquad$ Re $Z(p) > 0$ *for* Re $p > 0 \qquad\qquad\qquad (2.44)$

This is one of the most fundamental results of network theory. It establishes the *necessity* of the positive real condition for realizability. This condition will be seen to be *also sufficient*; but we cannot prove the *sufficiency* until we are in a position to demonstrate a synthesis procedure for an arbitrary positive real function. This objective is achieved in a later chapter, when our theory will be sufficiently developed to deal with this task. We now consider some important consequences and ramifications of the positive real condition.

Corollary 2.1 *If $Z(p)$ is a p.r.f., then $Y(p) = 1/Z(p)$ is also p.r.*

Corollary 2.2 *The sum of two positive real functions is also positive real; the difference, however, may not be positive real.*

To prove this result, let $Z_1(p)$ and $Z_2(p)$ be two p.r.f.s and consider their sum,

$$Z(p) = Z_1(p) + Z_2(p)$$

Hence

$$\text{Re } Z(p) = \text{Re } Z_1(p) + \text{Re } Z_2(p)$$

But $Z_1(p)$ and $Z_2(p)$ are real for p real, and Re $Z_1(p) > 0$, Re $Z_2(p) > 0$ for Re $p > 0$. Hence $Z(p)$ is real for p real and Re $Z(p) > 0$ for Re $p > 0$. Therefore $Z(p)$ is also p.r. However, for the difference

$$\bar{Z}(p) = Z_1(p) - Z_2(p)$$

$$\text{Re } \bar{Z}(p) = \text{Re } Z_1(p) - \text{Re } Z_2(p)$$

in which Re $\bar{Z}(p)$ cannot be guaranteed positive for Re $p > 0$. A further consequence is that if $Z(p)$ is a p.r.f., then $(Z(p) + k)$ and $kZ(p)$ are both p.r., where k is a positive constant.

Corollary 2.3 *A rational function* $Z(p)$ *satisfying the p.r. condition of Theorem 2.1, must be analytic in the open right half of the complex p-plane (i.e. all points satisfying* $\mathrm{Re}\, p > 0$*). Thus* $Z(p)$ *cannot have poles in the open right half-plane. Imaginary axis poles (i.e.* $\mathrm{Re}\, p = 0$*) are allowed, but if they occur they must be simple (of multiplicity 'one') with (real) positive residues.*

The above result is quite significant and its proof gives further insight into the properties of p.r. functions. For a definition and a brief discussion of analytic functions, see Appendix A.2. To show that a p.r.f. must be analytic in the open right half-plane, let p_0 be a pole of order n of the rational function $Z(p)$. In the neighbourhood of this pole, the function has a Laurent expansion of the form

$$Z(p) = \sum_{k=0}^{n} \frac{a_k}{(p - p_0)^k} + \sum_{k=0}^{\infty} b_k (p - p_0)^k \tag{2.45}$$

In polar form we may write

$$a_n = K_1 e^{j\theta_1}$$
$$(p - p_0) = K_2 e^{j\theta_2}$$

Thus, the real part of the dominant term of expansion (2.45) becomes

$$\mathrm{Re}\left\{ \frac{a_n}{(p - p_0)^n} \right\} = \frac{K_1}{K_2^n} \cos(\theta_1 - n\theta_2) \tag{2.46}$$

Now θ_1 is the angle which the residue a_n makes with the real axis and is fixed. As θ_2 is the angle of $(p - p_0)$ it can vary from 0 to 2π. But as this variation takes place, expression (2.46) changes sign $2n$ times. Thus in the vicinity of a pole p_0 in the open right half-plane, the real part of $Z(p)$ must also change sign $2n$ times. But this means that for any value of n, the real part of $Z(p)$ becomes negative, which contradicts the p.r. condition that $\mathrm{Re}\, Z(p)$ must remain positive for all $\sigma > 0$. It follows that $Z(p)$ cannot have any poles in the open right half-plane. However, if in (2.45) the pole p_0 is on the $j\omega$-axis, i.e. $\mathrm{Re}\, p_0 = 0$, then (2.46) is required to remain positive for $-(\pi/2) \leq n\theta_2 \leq \pi/2$. This is only possible for $n = 1$ and $\theta_1 = 0$, i.e. the $j\omega$-axis pole must be simple (of multiplicity $= 1$) and the residue a_n of the function at the pole must be real and positive.

A similar argument on the reciprocal of the impedance leads to the same conclusion about the poles of $Y(p)$, i.e. the zeros of $Z(p)$. Thus, a p.r.f. can have neither zeros nor poles in the open right half-plane ($\mathrm{Re}\, p > 0$). Imaginary axis poles ($\mathrm{Re}\, p = 0$) of either function must be simple with positive residues. Note that the points $p = 0$, ∞ are on the $j\omega$-axis. Also, a *positive* number is also *real* by implication, so it is enough to state that the residues are *positive*.

The right half-plane can be closed by adding its boundary: the entire $j\omega$-axis, *including* the point at ∞. Thus, a p.r.f is analytic in the closed right half-plane *except* at the poles on the boundary; see Fig. 2.6.

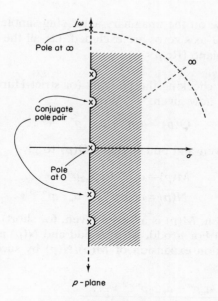

Fig. 2.6 Domain of analyticity of a
positive real function.

Corollary 2.4 *Write the positive real function $Z(p)$ as*

$$Z(p) = \frac{a_n p^n + a_{n-1} p^{n-1} + \ldots}{b_m p^m + b_{m-1} p^{m-1} + \ldots} \tag{2.47}$$

Then the highest powers of the numerator and denominator polynomials may differ by at most unity.

This is true since if n is greater than (or less than) m by more than unity, a multiple pole (or zero) at $\omega = \infty$ would result, which contradicts Corollary 2.3. The same statement can be made about the lowest powers of the numerator and denominator polynomials.

2.5 HURWITZ POLYNOMIALS

Our discussion so far has revealed that the locations of the poles and zeros of a p.r.f. play a central role in determining its properties. In particular, the exclusion of poles and zeros in the open right half-plane leads to the introduction of a class of polynomials known as *Hurwitz polynomials*. These are now defined and discussed.

Definition 2.2 A real polynomial $Q(p)$ (i.e. with real coefficients) is termed a Hurwitz polynomial if all its zeros lie in the closed left half-plane

(Re $p \leq 0$) with those on the imaginary axis being simple. $Q(p)$ is said to be *strictly* Hurwitz if $j\omega$-axis zeros are excluded, i.e. all the zeros occur only in the open left half-plane (Re $p < 0$).

A simple test for checking the Hurwitz (or strict Hurwitz) character of a polynomial $Q(p)$ is now given.[24,25] Write

$$Q(p) = a_n p^n + a_{n-1} p^{n-1} + \ldots \tag{2.48}$$

and separate the even and odd parts of $Q(p)$ to give

$$\begin{aligned} M(p) &= a_n p^n + a_{n-2} p^{n-2} + \ldots \\ N(p) &= a_{n-1} p^{n-1} + a_{n-3} p^{n-3} + \ldots \end{aligned} \tag{2.49}$$

where (a) For n even, $M(p)$ is all-even (even, for short) and $N(p)$ is all-odd (odd, for short). (b) For n odd, $M(p)$ is odd and $N(p)$ is even. Next, obtain the continued fraction expansion of $M(p)/N(p)$ by successive division and inversion as follows:

$$\begin{aligned} \frac{M(p)}{N(p)} &= \frac{a_n p^n + a_{n-2} p^{n-2} + \ldots}{a_{n-1} p^{n-1} + a_{n-3} p^{n-3} + \ldots} \\ &= \frac{a_n}{a_{n-1}} p + \frac{a'_{n-2} p^{n-2} + a'_{n-4} p^{n-4} + \ldots}{a_{n-1} p^{n-1} + a_{n-3} p^{n-3} + \ldots} = \alpha_1 p + \frac{M_1(p)}{N(p)} \end{aligned}$$

where $\alpha_1 p$ is the quotient resulting from the first division and $M_1(p)/N(p)$ is the remainder. Inverting this remainder and dividing again, we obtain

$$\frac{M(p)}{N(p)} = \alpha_1 p + \frac{1}{N(p)/M_1(p)} = \alpha_1 p + \frac{1}{\alpha_2 p + N_2(p)/M_1(p)}$$

This process of division and inversion is repeated until the function $M(p)/N(p)$ is exhausted. The result is the continued fraction expansion of $M(p)/N(p)$ (around $p = \infty$) of the form,

$$\frac{M(p)}{N(p)} = \alpha_1 p + \cfrac{1}{\alpha_2 p + \cfrac{1}{\alpha_3 p + \ldots}}$$

$$\phantom{\frac{M(p)}{N(p)} = } \ddots$$

$$\alpha_n p \tag{2.50}$$

Now, the necessary and sufficient conditions for $Q(p) = M(p) + N(p)$ to be *strictly* Hurwitz is that all the n quotients in the continued fraction expansion (2.50) be *strictly positive*, i.e.

$$\alpha_i > 0 \qquad i = 1, 2, \ldots, n \tag{2.51}$$

Next, suppose that all the α_is after α_k, say, are zero. Then the process in (2.50) terminates prematurely giving a zero remainder and indicting that $M(p)$ and $N(p)$ have a common factor. This is an even polynomial (because

M is odd and N is even or vice versa) which is identified as the numerator of the remainder in the division immediately preceding the zero-remainder step. This even common factor must be analysed separately. Consider a real even polynomial written as

$$E(p) = a_k p^k + a_{k-2} p^{k-2} + \ldots a_0$$

where k is even. The complex zeros of $E(p)$ must occur in quadruplets corresponding to factors of the form $(p + p_0)(p + p_0^*)(p - p_0)(p - p_0^*)$, otherwise the coefficients of $E(p)$ will not be real. Similarly a zero of $E(p)$ on the real axis must give rise to a factor of the form $(p^2 - \sigma_0^2)$. Now, the definition of a Hurwitz polynomial excludes zeros in the open right half-plane. But complex zeros of a real even polynomial occur in quadruplets, and real-axis ones occur at $p = \pm\sigma_0$. Hence, excluding open right half-plane zeros of a real even polynomial immediately excludes their images in the open left half-plane. On the other hand, purely imaginary zeros of $E(p)$ occur in conjugate pairs of the form $(p^2 + \omega_0^2)$, where ω_0 is real, and these factors are allowed by Definition 2.2 provided they are simple. Therefore, in order for a real even polynomial to be Hurwitz, it must have only simple zeros on the imaginary axis.

A simple procedure for determining whether an even polynomial $E(p)$ has only simple purely imaginary zeros, is to differentiate $E(p)$ to obtain $E'(p) = dE(p)/dp$, then the required condition is that $E(p) + E'(p)$ be strictly Hurwitz, i.e. the continued fraction expansion of $E(p)/E'(p)$ yields all positive quotients.

Example 2.1   Test the following polynomial for Hurwitz character:

$$Q(p) = p^3 + 6p^2 + 11p + 6$$

Solution   Separate $Q(p)$ into its odd and even parts,

$$M(p) = p^3 + 11p, \qquad N(p) = 6p^2 + 6$$

The continued fraction expansion of $M(p)/N(p)$ is obtained as follows

$$
\begin{array}{r|l}
 & \frac{1}{6}p \\
6p^2+6 & p^3+11p \\
 & p^3+p \\
\hline
 & \quad\quad \frac{3}{5}p \\
10p & 6p^2+6 \\
 & 6p^2 \\
\hline
 & \quad\quad \frac{5}{3}p \\
6 & 10p \\
 & 10p \\
\hline
 & 0
\end{array}
$$

The process terminates after three steps (the degree of the polynomial) yielding all positive quotients. Hence, the polynomial is strictly Hurwitz.

Example 2.2　Test the following polynomial for Hurwitz character

$$Q(p) = p^8 + 3p^7 + 10p^6 + 24p^5 + 35p^4 + 57p^3 + 50p^2 + 36p + 24$$

Solution　Write

$$M(p) = p^8 + 10p^6 + 35p^4 + 50p^2 + 24$$
$$N(p) = 3p^7 + 24p^5 + 57p^3 + 36p$$

Performing the continued fraction expansion of $M(p)/N(p)$ we have

$$
\begin{array}{r}
\frac{1}{3}p \\[4pt]
3p^7 + 24p^5 + 57p^3 + 36p \;\overline{\big)\; p^8 + 10p^6 + 35p^4 + 50p^2 + 24} \\
p^8 + 8p^6 + 19p^4 + 12p^2
\end{array}
$$

$$
\begin{array}{r}
\frac{3}{2}p \\[4pt]
2p^6 + 16p^4 + 38p^2 + 24 \;\overline{\big)\; 3p^7 + 24p^5 + 57p^3 + 36p} \\
3p^7 + 24p^5 + 57p^3 + 36p \\ \hline
0 \qquad 0 \qquad 0 \qquad 0
\end{array}
$$

The first two quotients are positive, but the process has terminated after two steps, giving a zero remainder. Thus, $M(p)$ and $N(p)$ have the even polynomial $(2p^6 + 16p^4 + 38p^2 + 24)$ as a common factor. Therefore $Q(p)$ has an even factor,

$$E(p) = p^6 + 8p^4 + 19p^2 + 12$$

It is now required to determine whether $E(p)$ has only simple pure imaginary zeros. First differentiate $E(p)$ to obtain

$$E'(p) = 6p^5 + 32p^3 + 38p$$

Then the continued fraction expansion of $E(p)/E'(p)$ is found. This gives six strictly positive partial quotients:

$$\tfrac{1}{6}p, \quad \tfrac{9}{4}p, \quad \tfrac{16}{21}p, \quad \tfrac{49}{60}p, \quad \tfrac{25}{7}p, \quad \text{and} \quad \tfrac{1}{10}p$$

Hence the polynomial $E'(p) + E(p)$ is strictly Hurwitz. It follows that the zeros of $E(p)$ are simple and on the imaginary axis. Adding this result to the fact that the first two quotients in the expansion of $M(p)/N(p)$ are positive, we conclude that the given polynomial is Hurwitz.

Now, Corollary 2.3 and the discussion which followed, showed that a p.r.f. must be devoid of poles and zeros in the open right half-plane. Combining this with Definition 2.2 of a Hurwitz polynomial we obtain the following result.

Corollary 2.5 *Let a p.r.f.* $Z(p)$ *be written as*

$$Z(p) = \frac{Q(p)}{D(p)}$$

where $Q(p)$ *and* $D(p)$ *are relatively prime polynomials* (*i.e. they have no common factors*). *Then both* $Q(p)$ *and* $D(p)$ *are Hurwitz polynomials.*

2.6 AN ALTERNATIVE SET OF CONDITIONS FOR POSITIVE REAL CHARACTER

In testing a rational function for positive real character according to Theorem 2.1, it is easy to check its realness (i.e. $Z(p)$ is real for p real). This only requires that the numerator and denominator be polynomials with real coefficients. A consequence of this is that the zeros of each polynomial must be symmetrically distributed with respect to the real axis of the p-plane. This is because complex zeros must occur in conjugate pairs in order for the factor corresponding to each pair to have real coefficients. However, checking condition (ii) of Theorem 2.1 requires that the function be tested for positive real part over the entire open right half-plane; which can be a very tedious task. This provides a motivation for deriving a more convenient set of conditions for a rational function to be positive real. These can be obtained by combining Theorem 2.1 with a theorem in complex variables known as the *minimum real-part theorem*, which is given in Appendix A.2. The required result is stated below without proof.

Corollary 2.6 *The necessary and sufficient conditions for a real rational function* $Z(p)$ *to be positive real are*

(a) $Z(p)$ *is analytic in* $\operatorname{Re} p > 0$.
(b) $\operatorname{Re} Z(j\omega) \geq 0$ *for* $-\infty \leq \omega \leq \infty$, *i.e. along the entire* $j\omega$*-axis.*
(c) *Any* $j\omega$*-axis poles, including a possible pole at infinity, must all be simple and have* (*real*) *positive residues.*

The above result allows us to check the p.r. character of a function by examining the behaviour of its real-part on the imaginary axis only instead of over the entire right half-plane.

2.7 EXTRACTION OF IMAGINARY-AXIS POLES FROM A POSITIVE REAL FUNCTION

So far, we have discussed the poles and zeros of a p.r.f. merely as abstract concepts; thus restricting their location to the left half-plane or the imaginary axis. We now unveil some of the more concrete physical aspects of positive real functions. In particular, we give an interpretation, in terms of a physical network, of imaginary-axis poles and zeros of a p.r.f.

Let $j\omega_i$, $i = 1, 2, \ldots, m$ be simple poles on the imaginary axis, of a p.r.f. $Z(p)$ satisfying the conditions of Theorem 2.1 or Corollary 2.6. The function may, in addition have a simple pole at $p = 0$ and a simple pole at $p = \infty$. Thus, $Z(p)$ may be expressed in the form

$$Z(p) = \frac{k_0}{p} + \sum_{i=1}^{m} \frac{2k_i p}{p^2 + \omega_i^2} + k_\infty p + Z_1(p) \tag{2.52}$$

where the constituent terms are interpreted as follows:

(a) A term k_0/p results from a possible pole at $p = 0$.
(b) A term $k_\infty p$ results from a possible pole at $p = \infty$.
(c) A term $2k_i p/(p^2 + \omega_i^2)$ results from a pair of conjugate poles on the $j\omega$-axis,

$$\frac{k_i}{p + j\omega_i} + \frac{\bar{k}_i}{p - j\omega_i} = \frac{(k_i + \bar{k}_i)p - j(k_i - \bar{k}_i)\omega_i}{p^2 + \omega_i^2} \tag{2.53}$$

and in order for the terms corresponding to the conjugate pair to combine giving a real term (remember Z must be real for p real), then the two residues k_i and $\bar{k}_i$ in (2.53) must be equal, i.e. $k_i = \bar{k}_i$ and we obtain the typical term in the summation of (2.52).
(d) $Z_1(p)$ is a function devoid of $j\omega$-axis poles, including 0 and ∞.

The residues of the function at the poles may be calculated from the following expressions.

$$k_0 = \{pZ(p)\}_{p=0} \tag{2.54}$$

$$k_\infty = \lim_{p \to \infty} \left\{ \frac{Z(p)}{p} \right\} \tag{2.55}$$

$$2k_i = \left\{ \frac{(p^2 + \omega_i^2)}{p} Z(p) \right\}_{p=j\omega_i} \tag{2.56}$$

Now from the p.r. conditions in Theorem 2.1 or Corollary 2.6, and the expansion of the impedance in (2.52) we have the following result.

Corollary 2.7 *Let $Z(p)$ be a p.r.f. possessing the expansion* (2.52). *Then $Z_1(p)$, the impedance which remains after subtracting from $Z(p)$ all (or some) of the terms resulting from $j\omega$-axis poles including those at 0 and ∞, is also p.r. Thus,*

$$Z_1(p) = Z(p) - \frac{k_0}{p} - \sum_{i=1}^{m} \frac{2k_i p}{p^2 + \omega_i^2} - k_\infty p \tag{2.57}$$

is a p.r.f.

We speak of the process in (2.57) as the 'extraction' of $j\omega$-axis poles from a given $Z(p)$. It is easy to see that $Z_1(p)$ is also p.r. Note that since $Z(p)$ is p.r., then from (2.57), $Z_1(p)$ is real for p real, and is also analytic in the open

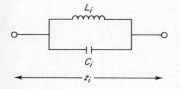

Fig. 2.7 A circuit for the realization of a typical term in the summation of (2.57).

right half-plane. Moreover,

$$\text{Re } Z_1(j\omega) = \text{Re } Z(j\omega) \geq 0 \qquad (2.58)$$

Therefore, $Z_1(j\omega)$ satisfies Corollary 2.6 and is a p.r.f. This means that a p.r.f. remains p.r. after the extraction of all (or some) of its poles on the imaginary axis (those at 0 and ∞ included). The degree of the function is diminished by the number of extracted poles.

Now, what does extracting a $j\omega$-axis pole physically mean? We recognize that in (2.57), the term k_0/p is the impedance of a capacitor, $k_\infty p$ is the impedance of an inductor, and a typical term in the summation $2k_ip/(p^2 + \omega_i^2)$ is the impedance of a parallel L_iC_i as shown in Fig. 2.7,

$$z_i = \frac{(1/C_i)p}{p^2 + 1/L_iC_i} \qquad (2.59)$$

Since these terms are extracted from an impedance, the corresponding circuits are placed in series. We are thus led to the network interpretation of the process of $j\omega$-axis pole extraction shown in Fig. 2.8. The element values are obtained by comparing the various terms in (2.57) with the impedance of a capacitor, an inductor, and that of a parallel L_iC_i as given by (2.59). Hence,

$$\begin{aligned} C_0 &= 1/k_0, & L_\infty &= k_\infty \\ C_i &= 1/2k_i, & L_i &= 2k_i/\omega_i^2 \end{aligned} \qquad (2.60)$$

Clearly, the process of $j\omega$-axis pole extraction can also be applied to a p.r. admittance $Y(p)$. Instead of (2.57) we have

$$Y_1(p) = Y(p) - \frac{h_0}{p} - \sum_{i=1}^{r} \frac{2h_ip}{p^2 + \omega_i^2} - h_\infty p \qquad (2.61)$$

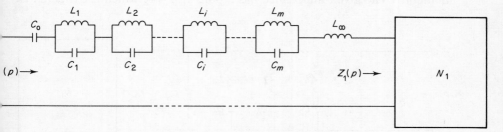

Fig. 2.8 Extraction of $j\omega$-axis poles from a positive real impedance.

Fig. 2.9 A circuit for the realization of a typical term in the summation of (2.61).

which is also p.r. Here h_0/p is the admittance of an inductor, $h_\infty p$ that of a capacitor, and $2h_i p/(p^2 + \omega_i^2)$ is the admittance of a series $L_i C_i$ as shown in Fig. 2.9 which is given by

$$y_i = \frac{(1/L_i)p}{p^2 + 1/L_i C_i} \tag{2.62}$$

Therefore, we have the network interpretation of (2.61) shown in Fig. 2.10, with the element values

$$
\begin{aligned}
L_0 &= 1/h_0, & C_\infty &= h_\infty \\
L_i &= 1/2h_i, & C_i &= 2h_i/\omega_i^2
\end{aligned}
\tag{2.63}
$$

Definition 2.3 A positive real impedance which is devoid of $j\omega$-axis poles (including 0 and ∞) is termed minimum reactance, whereas a positive real admittance devoid of $j\omega$-axis poles (including 0 and ∞) is termed minimum susceptance.

Thus, if we have a p.r.f. and extract all $j\omega$-axis poles (0 and ∞ included) from the function and its reciprocal, the remaining function is minimum reactance and minimum susceptance.

Example 2.3 Extract all $j\omega$-axis poles (0 and ∞ included) from the following p.r.f. and its reciprocal, giving the network interpretation.

$$Z(p) = \frac{3p^5 + 13p^4 + 14p^3 + 18p^2 + 6p + 4}{3p^4 + 13p^3 + 8p^2 + 4p + 2}$$

Solution The given impedance has a pole at $p = \infty$, which is extracted as

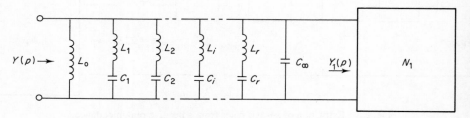

Fig. 2.10 Extraction of $j\omega$-axis poles from a positive real admittance.

a series inductor of value,

$$L_1 = \lim_{p \to \infty} \left\{ \frac{Z(p)}{p} \right\} = 1 \, \text{H}$$

The remainder is given by

$$Z_1(p) = Z(p) - p$$

$$= \frac{6p^3 + 14p^2 + 4p + 4}{3p^4 + 13p^3 + 8p^2 + 4p + 2}$$

Therefore, $Y_1(p) = 1/Z_1(p)$ has a pole at $p = \infty$, which is extracted as a shunt capacitor of value

$$C_2 = \lim_{p \to \infty} \left\{ \frac{Y_1(p)}{p} \right\} = \tfrac{1}{2} \, \text{F}$$

The remainder admittance is given by

$$Y_2(p) = Y_1(p) - \frac{p}{2}$$

$$= \frac{3p^3 + 3p^2 + p + 1}{3p^3 + 7p^2 + 2p + 2}$$

or

$$Z_2(p) = \frac{3p^3 + 7p^2 + 2p + 2}{(3p^2 + 1)(p + 1)}$$

which has a pole at $p = \pm j1\sqrt{3}$. This is extracted as a parallel LC in series. The residue k_2 of $Z_2(p)$ at this pole is calculated using (2.56),

$$2k_2 = \left\{ \frac{(3p^2 + 1)}{3p} Z_2(p) \right\}_{p=j1/\sqrt{3}} = \tfrac{1}{3}$$

Extraction of this pole leaves a remainder,

$$Z_3(p) = \frac{3p^3 + 7p^2 + 2p + 2}{3p^3 + 3p^2 + p + 1} - \frac{\frac{1}{3}p}{(p^2 + \frac{1}{3})}$$

$$= \frac{(p + 2)}{(p + 1)}$$

which is minimum reactance and minimum susceptance. The network interpretation of the above procedure is shown in Fig. 2.11.

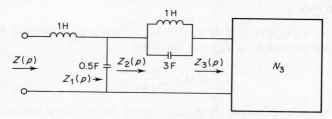

Fig. 2.11   Network of Example 2.3.

2.8 THE BOUNDED REAL CONCEPT

We now introduce a very useful concept which is closely related to the positive real condition. It is used here to derive some interesting properties of p.r. functions, and to show that a one-port may be characerized by an alternative function as well as its impedance or admittance. Later, it will be seen that this concept is also indispensable in the description of two-port and n-port networks.

Definition 2.4 A function $S(p)$ is aid to be bounded real if it satisfies

(a) $S(p)$ is real for p real
(b) $|S(p)| < 1$ in Re $p > 0$

Alternatively, condition (b) may be replaced by

(c) $S(p)$ is analytic in Re $p > 0$
(d) $|S(j\omega)| \leq 1,\qquad -\infty \leq \omega \leq \infty$

It can be shown that (c) and (d) together are equivalent to (b) using a result in complex variables known as the *maximum modulus theorem* which is given in Appendix A.2. This states that if a function is analytic within and on the contour of a given region, then the modulus of the function has its maximum on the contour and not at any interior point. Applying this theorem to a function $S(p)$ satisfying conditions (c) and (d) above, we see that $|S(p)|$ is required to be less than or equal to unity on the $j\omega$-axis which is part of the boundary of the closed right half-plane. Therefore $S(p)$ is bounded by unity at all interior points and hence satisfies (b). Combining Definition 2.4 with Definition 2.2 of a Hurwitz polynomial, we obtain the following result.

Theorem 2.2 *For a function* $S(p) = n(p)/d(p)$ *to be bounded real, the necessary and sufficient conditions are*:

(i) $S(p)$ *is real for p real*
(ii) $d(p)$ *is strictly Hurwitz, i.e.* $d(p) \neq 0$ *in* Re $p \geq 0$
(iii) $|S(j\omega)| \leq 1,\qquad -\infty \leq \omega \leq \infty$

2.9 RELATING THE POSITIVE REAL AND BOUNDED REAL CONDITIONS

Let $Z(p)$ be a positive real function. It is shown in Appendix A.3 that a bilinear transformation of $Z(p)$ of the form

$$S(p) = \frac{Z(p) - 1}{Z(p) + 1} \tag{2.64}$$

is a bounded real function. It is also shown in the same appendix that a bilinear transformation of a bounded real function $S(p)$ of the form

$$Z(p) = \frac{1+S(p)}{1-S(p)} \tag{2.65}$$

is a positive real function. Hence, given a positive real function $Z(p)$, a bounded real function may be formed using (2.64). Conversely, given a bounded real function $S(p)$, one can form a positive real function using (2.65). But if $Z(p)$ is the driving-point impedance of a one-port, what is $S(p)$ as obtained from (2.64)? We now show that the answer to this question follows naturally from an alternative description of the one-port.

2.10 THE SCATTERING DESCRIPTION OF A ONE-PORT

Consider the generator of e.m.f. $V_g(j\omega)$ and internal resistance r_g when connected to a resistive load R as shown in Fig. 2.12(a). The active power dissipated in the load is

$$W(j\omega) = \frac{|V_g(j\omega)|^2 R}{(R+r_g)^2} \tag{2.66}$$

and this is maximum for $R = r_g$, i.e.

$$W_{max} = \frac{|V_g(j\omega)|^2}{4r_g} \tag{2.67}$$

Now, let the generator be connected to a one-port network N whose driving-point impedance is $Z(p)$ as shown in Fig. 2.12(b). For complex frequencies we have

$$I(p) = \frac{V_g(p)}{r_g + Z(p)} \tag{2.68}$$

and for $Z(p) = r_g$,

$$I_0 = \frac{V_g(p)}{2r_g} \tag{2.69}$$

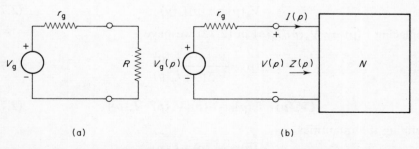

(a)   (b)

Fig. 2.12 Pertinent to the scattering description of a one-port. (a) A resistive source connected to a resistor R, (b) the source connected to a passive one-port N.

From (2.68) and (2.69),

$$I_0(p) - I(p) = \frac{\{Z(p) - r_g\} V_g(p)}{2r_g\{Z(p) + r_g\}} \tag{2.70}$$

The relative difference between (2.70) and (2.69) is then given by

$$\frac{I_0(p) - I(p)}{I_0(p)} = \frac{Z(p) - r_g}{Z(p) + r_g}$$

$$= S(p) \tag{2.71}$$

which is called the *reflection coefficient* of $Z(p)$ relative to r_g; the reason for using this term will be apparent shortly. From (2.71) we have

$$S(p) = \frac{Z(p) - r_g}{Z(p) + r_g}$$

$$= \frac{z(p) - 1}{z(p) + 1} \tag{2.72}$$

where $z(p) = Z(p)/r_g$ is the normalized impedance of the one-port.

It now follows from (2.64) and (2.72) that if $z(p)$ is a p.r.f. then $S(p)$ is bounded real. We also have the inverse relation

$$z(p) = \frac{1 + S(p)}{1 - S(p)} \tag{2.73}$$

or

$$Z(p) = r_g \frac{1 + S(p)}{1 - S(p)} \tag{2.74}$$

Thus, if $Z(p)$ in (2.72) is the input impedance of the one-port, then $S(p)$ is its reflection coefficient. We now justify the use of the term *reflection*. Consider again the one-port N in Fig. 2.12(b). In terms of the *normalized* quantities,

$$V_n(p) = \frac{V(p)}{\sqrt{r_g}}, \quad I_n(p) = \sqrt{r_g}\, I(p), \quad z(p) = \frac{Z(p)}{r_g} \tag{2.75}$$

the one-port is described by its impedance equation

$$V_n(p) = z(p) I_n(p) \tag{2.76}$$

Replacing $z(p)$ by $V_n(p)/I_n(p)$ in (2.72) we have

$$S(p) = \frac{V_n(p) - I_n(p)}{V_n(p) + I_n(p)} \tag{2.77}$$

or

$$\{V_n(p) - I_n(p)\} = S(p)\{V_n(p) + I_n(p)\} \tag{2.78}$$

Defining the quantities

$$\begin{aligned} \alpha(p) &= V_n(p) + I_n(p) \\ \beta(p) &= V_n(p) - I_n(p) \end{aligned} \tag{2.79}$$

we can write

$$\beta(p) = S(p)\alpha(p) \tag{2.80}$$

The above expression is an alternative way of relating the variables $V_n(p)$ and $I_n(p)$ as compared with the impedance equation (2.76). Therefore (2.80) characterizes the one-port just as well as (2.76).

Now for the network in Fig. 2.12(b) we may write

$$V(p) = Z(p)I(p)$$

and

$$V(p) = V_g(p) - r_g I(p) \tag{2.81}$$

In terms of the normalized variables, the above expressions become

$$V_n(p) = z(p)I_n(p)$$

and

$$V_n(p) = V_{gn}(p) - I_n(p) \tag{2.82}$$

where $V_{gn}(p) = V_g(p)/\sqrt{r_g}$. Comparing (2.79) with (2.82) we have

$$V_{gn}(p) \equiv \alpha(p) \tag{2.83}$$

so that $\alpha(p)$ is the normalized generator voltage. From (2.71) the actual current in the network is

$$I(p) = I_0(p) - S(p)I_0(p) \tag{2.84}$$

which is the superposition of the current at matching: $I_0(p)$, and a reflected current $S(p)I_0(p)$ (flowing in opposite direction) produced by a fictitious generator located in the one-port. The normalized e.m.f. of the fictitious generator is given by

$$S(p)V_{gn}(p) = S(p)\alpha(p)$$
$$= \beta(p) \tag{2.85}$$

Therefore, $\beta(p)$ is called the reflected signal. The incident signal $\alpha(p)$ and the reflected signal $\beta(p)$ are related by the reflection coefficient $S(p)$ as in (2.80). This is illustrated in Fig. 2.13, and is a simple example of the *scattering description* of networks.

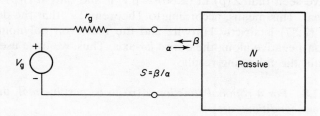

Fig. 2.13 The scattering description of a passive one-port.

2.11 TESTING A FUNCTION FOR POSITIVE REAL CHARACTER

An immediate application of the bounded real concept is to derive an alternative set of necessary and sufficient conditions for a function to be positive real. These will be much easier to verify than the conditions of Theorem 2.1 or Corollary 2.6. First note that condition (b) of Corollary 2.6 namely: Re $Z(j\omega) \geq 0$ is equivalent to $|S(j\omega)| \leq 1$ in the bilinear transformation (2.64). Thus we can state that a given function $Z(p)$ is positive real if and only if the corresponding reflection coefficient (relative to an arbitrary positive resistor r_g taken to be 1 Ω for convenience), is bounded real. Write

$$Z(p) = \frac{N(p)}{D(p)} \qquad (2.86)$$

and the reflection coefficient of the one-port relative to $r_g = 1 \Omega$ is given by

$$\begin{aligned} S(p) &= \frac{Z(p) - 1}{Z(p) + 1} \\ &= \frac{N(p) - D(p)}{N(p) + D(p)} \end{aligned} \qquad (2.87)$$

where it is assumed that $N(p)$ and $D(p)$ are relatively prime polynomials. From (2.87) we have

$$1 - S(p)S(-p) = 2\frac{N(p)D(-p) + N(-p)D(p)}{\{D(p) + N(p)\}\{D(-p) + N(-p)\}} \qquad (2.88)$$

or

$$1 - SS_* = \frac{2u(p)}{(D + N)(D_* + N_*)} \qquad (2.89)$$

where the lower asterisk denotes replacing p by $-p$ and

$$u(p) = ND_* + N_*D \qquad (2.90)$$

is a real even polynomial (because $u(p) = u(-p)$). Evaluating (2.89) at $p = j\omega$ one obtains,

$$1 - |S(j\omega)|^2 = \frac{2\hat{u}(\omega^2)}{|D(j\omega) + N(j\omega)|^2} \qquad (2.91)$$

But we have seen that $Z(p)$ in (2.86) is p.r. if and only if $S(p)$ in (2.87) is bounded real. This means, according to Theorem 2.2, that the denominator of $S(p)$ in (2.87) is strictly Hurwitz and the real even polynomial $u(p)$ in (2.90) is non-negative along the entire $j\omega$-axis. Thus, we have used Theorem 2.2 to obtain the following result.

Corollary 2.8 *For a rational function $Z(p)$ to be positive real, the necessary and sufficient conditions are:*

(a) $Z(p) = N(p)/D(p)$ *is real for p real*

(b) $N(p) + D(p)$ *is strictly Hurwitz*

(c) *The real even polynomial*

$$u(p) = ND_* + N_*D$$

is non-negative along the entire $j\omega$-axis, i.e.

$$\{u(p)\}_{p=j\omega} = \hat{u}(\omega^2) \geq 0 \qquad -\infty \leq \omega \leq \infty$$

The above conditions form a convenient basis for testing a function for positive real character. Condition (a) is easily checked by ascertaining that $N(p)$ and $D(p)$ are real polynomials (i.e. with real coefficients). Condition (b) can be checked using the test given in Section 2.5 for the strict Hurwitz character of a polynomial. Therefore we are left with the problem of checking condition (c) of Corollary 2.8.

Consider the typical factors of *any real even* polynomial $\hat{u}(\omega^2)$. These are as follows:

(i) Factors corresponding to complex zeros are automatically of the form $(\omega^2 + p_0^2)(\omega^2 + p_0^{*2})$ in order to preserve the realness of the polynomial coefficients. These factors combine to form terms which are positive for all ω.

(ii) Factors of the form $(\omega^2 + \alpha^2)$, where α is real, clearly remain positive for all ω.

(iii) Factors of the form $(\omega^2 - \alpha^2)$, where α is real, clearly become negative for some ω values, unless they occur with even multiplicity. Thus each factor of this type must be of the form $(\omega^2 - \alpha^2)^{2k}$ in order for it to be non-negative for all ω.

From the above discussion of the possible factors of $\hat{u}(\omega^2)$ it is clear that in order for $\hat{u}(\omega^2)$ to be non-negative for all ω, it is necessary and sufficient that all its real positive ω^2 zeros be of even multiplicity. Letting $\omega^2 = x$, this means that either (1) $\hat{u}(x)$ is devoid of real zeros in the interval $0 < x < \infty$, or (2) any real positive zeros of $\hat{u}(x)$ must occur with even multiplicity.

Thus, the problem reduces to one of determining whether $u(x)$ has any positive zeros; if they occur they must be of even multiplicity. This property can be checked using a 'brute force' approach by factorizing the polynomial and examining the location and multiplicity of its zeros. In these days of high speed computational algorithms and the universal availability of digital computers (the reader might have one at home!) this presents no real problem. However, there exists a relatively straightforward algorithm for testing a polynomial for non-negativitiy; this can be found in many references.[24-26]

Example 2.4 Test the following functions for positive real character

(a)

$$Z(p) = \frac{(p+2)^2}{p^2 + 4}$$

44

(b)
$$Z(p) = \frac{p^2 + 2p + 2}{p^2 + 2p + 1}$$

Solution   The conditions of Corollary 2.8 are applied to each of the given functions.

(a)
$$Z(p) = \frac{p^2 + 4p + 4}{p^2 + 4}$$

Clearly $Z(p)$ is real for p real. Also the sum of numerator and denominator is $2p^2 + 4p + 8$, which upon use of expansion (2.50) is shown to be strictly Hurwitz. The polynomial $u(p)$ in Condition (c) of Corollary 2.8 is given by

$$u(p) = (p^2 + 4p + 4)(p^2 + 4) + (p^2 - 4p + 4)(p^2 + 4)$$
$$= 2(p^2 + 4)^2$$

or

$$\{u(p)\}_{p=j\omega} = \hat{u}(\omega^2) = 2(4 - \omega^2)^2$$

which has a real and positive ω^2 zero with even multiplicity, therefore it is non-negative for all ω. Hence, $Z(p)$ is p.r, satisfying (a) to (c) of Corollary 2.8.

(b)
$$Z(p) = \frac{p^2 + 2p + 2}{p^2 + 2p + 1}$$

Again, condition (a) of Corollary 2.8 is satisfied since $Z(p)$ is real for p real. Also the sum of numerator and denominator is $2p^2 + 4p + 3$ which is strictly Hurwitz, thus condition (b) of Corollary 2.8 is also satisfied. Next evaluate the polynomial $u(p)$ in condition (c) of Corollary 2.8,

$$u(p) = (p^2 + 2p + 2)(p^2 - 2p + 1) + (p^2 - 2p + 2)(p^2 + 2p + 1)$$
$$= 2p^4 - 2p^2 + 4$$

and

$$\hat{u}(\omega^2) = 2\omega^4 + 2\omega^2 + 4$$

which is clearly non-negative for all ω. Hence, the function also satisfies (c) of Corollary 2.8. Therefore $Z(p)$ is p.r.

We end this section by giving a useful property of positive real functions, which follows from Corollary 2.8. The proof is left to the reader.

Corollary 2.9   *Let $Z(p)$ be a p.r.f. and write*

$$Z(p) = \frac{m_1(p) + n_1(p)}{m_2(p) + n_2(p)}$$

where $m_{1,2}$ are even, and $n_{1,2}$ are odd polynomials. Then the function

$$Z_1(p) = \frac{m_1(p) + n_2(p)}{m_2(p) + n_1(p)}$$

is also p.r. In words, a p.r.f. remains p.r. if we interchange the even parts (or odd parts) of its numerator and denominator.

2.12 RECAPITULATION

We have seen that the driving-point impedance of a passive, lumped, linear, and time-invariant network, must be a positive real function (p.r.f.), thus establishing the necessity of the p.r. condition for realizability. Although our interest is mainly concentrated in reciprocal networks containing the conventional elements, we have shown that if ideal gyrators are allowed to exist in the network, the p.r. condition is still satisfied. Testing a function for p.r. character is most conveniently accomplished by the use of Corollary 2.8. We also introduced the bounded real condition which is closely related to the p.r. condition. It was shown that a one-port may be described by its reflection coefficient, which is related to its driving-point impedance by a bilinear transformation of the form (2.64) or (2.65).

PROBLEMS

2.1 Test the following polynomials for Hurwitz character,

(a) $p^3 + 6p^2 + 10p + 8$

(b) $p^4 + 2p^3 + 3p^2 + 4p + 3$

(c) $p^5 + 6p^4 + 12p^3 + 12p^2 + 11p + 5$

(d) $p^5 + 7p^4 + 16p^3 + 18p^2 + 10p + 2$

(e) $p^6 + 7p^4 + 14p^2 + 8$

(f) $p^8 + 1$

2.2 Test the following polynomials for non-negativity for all real ω. (The solution may require the use of a computer.)

(a) $\omega^{10} + \omega^8 + 3\omega^6 - 3\omega^4 + 4\omega^2 + 10$

(b) $\omega^{10} - 3\omega^8 + 4\omega^6 - 4\omega^2 + 4$

(c) $\omega^{10} + 5\omega^8 + 15\omega^6 - 25\omega^4 + 24\omega^2 + 10$

(d) $\omega^{10} + \omega^8 + 3\omega^6 - 3\omega^4 + 4\omega^2 + 10$

2.3 Show that if $Z(p)$ is a p.r.f., then $Z(k/p)$ is also p.r., where k is a positive constant.

2.4 Show that if $Z(p)$ and $W(p)$ are both p.r., then $Z(W)$ is also p.r.

2.5 Test the following functions for p.r. character

(a) $\dfrac{1}{p^2 + 3p + 2}$

(b) $\dfrac{p^2 + 7p + 12}{p^2 + 3p + 2}$

(c) $\dfrac{p^2 + p + 2}{2p^2 + 2p + 1}$

(d) $\dfrac{2p^2 + 2p + 1}{2p^3 + 2p^2 + 2p + 1}$

(e) $\dfrac{2p^3+3p^2+6p+1}{p^2+p+1}$ (f) $\dfrac{p^4+4p^3+3p^2+4p+1}{p^4+p^3+3p^2+p+1}$

(g) $\dfrac{p(p+3)(p+5)}{(p+1)(p+4)}$ (h) $\dfrac{p^3+5p^2+4p}{p^4+8p^2+15}$

2.6 Let $Z(p)$ be a rational p.r.f. and let k be a positive constant.

(a) Show that the function

$$Z_1(p) = \frac{kZ(p)-pZ(k)}{kZ(p)-pZ(p)}$$

$$= \frac{N_1(p)}{D_1(p)}$$

is also p.r.

(b) Show that N_1 and D_1 have a common factor $(p-k)$.

(c) Show that N_1 and D_1 will have a common factor (p^2-k^2) if $p=k$ is a zero of the even-part of $Z(p)$, i.e. if

$$\{Z(p)+Z_*(p)\}_{p=k} = 0$$

Chapter 3
Synthesis of Lossless One-ports

3.1 INTRODUCTION

The most general type of passive one-port contains all the conventional elements namely: resistors, capacitors, inductors (susceptible to mutual coupling), and ideal transformers. However, many important applications require the use of only two kinds of elements. If only inductors and capacitors are allowed then the network is of the lossless (LC) type. If only resistors and capacitors are allowed, the network is of the RC type. Similarly an RL network contains resistors and inductors only.

The driving-point impedance of a one-port formed using two kinds of elements only must be a positive real function. Moreover, in each case the impedance is a special case of a p.r.f. and, therefore, satisfies further constraints in addition to the p.r. condition.

Of the three types of one-port containing two kinds of elements, RL ones are of no practical importance. RC and LC networks are both important from the theoretical as well as the practical viewpoints. However, we shall consider only lossless (LC) one-ports, for two reasons. First, these are the networks of direct relevance to the material in this book. Secondly, the properties and synthesis techniques of RC one-ports can be derived from those of LC one-ports by a simple transformation. The reader who is interested in RC one-ports is referred to the textbooks which deal with this topic.[24,25]

3.2 REACTANCE (FOSTER) FUNCTIONS

We know that the driving-point impedance of any passive one-port must be a p.r.f. If, in addition, the network is lossless, there is no dissipation. Therefore expression (2.39) evaluated with $W_d = 0$, on the $j\omega$-axis ($\sigma = 0$) gives for a lossless one-port

$$Z(j\omega) + Z(-j\omega) = 0 \qquad \text{for all } \omega \qquad (3.1)$$

i.e.

$$\{\text{Ev } Z(p)\}_{p=j\omega} = \tfrac{1}{2}\{Z(p) + Z_*(p)\}_{p=j\omega}$$
$$= 0 \qquad (3.2)$$

But since $Z(p)$ is analytic, so is $Z_*(p)$. Therefore the sum in (3.2) is also analytic, and cannot vanish at an infinite number of points (all $j\omega$) without vanishing identically. Thus

$$\text{Ev } Z(p) = \tfrac{1}{2}\{Z(p) + Z_*(p)\}$$
$$= 0 \qquad \text{for all } p \qquad (3.3)$$

so that $Z(p)$ is an odd rational function, i.e. it is the ratio of odd-to-even or even-to-odd polynomials. Note that the above argument, from points on the $j\omega$-axis to all-points in the p-plane, is an example of a very powerful mathematical technique known as *analytic continuation*.

Definition 3.1 A positive real impedance $Z(p)$ which is an odd rational function (i.e. satisfying (3.3)) is called a reactance function or a Foster function.

3.3 FOSTER'S CANONIC FORMS

According to Definition 3.1 and the preceding discussion, in order for a given function $Z(p)$ to be realizable as the driving-point impedance of a lossless one-port, it is *necessary* that $Z(p)$ be a reactance (or Foster) function. We now develop a number of further properties of such functions leading to a realization in terms of a lossless one-port, thus establishing the *sufficiency* of the reactance condition.

Now, since a reactance function $Z(p)$ is the ratio of even-to-odd or odd-to-even polynomials, their degrees must differ by exactly unity by Corollary 2.4. Therefore the points $p = 0$ and $p = \infty$ are always critical, i.e. a pole or a zero. Moreover, the numerator and denominator of $Z(p)$ must by Hurwitz polynomials by Corollary 2.5. But we have seen that all the zeros of a real even (or odd) Hurwitz polynomial must be simple and on the $j\omega$-axis. Thus, all the poles and zeros of a reactance function are simple and lie on the $j\omega$-axis. Furthermore, according to Corollary 2.3 the residues of the function (and its reciprocal) at the poles are real and positive. Therefore, we may express the driving-point impedance of a lossless one-port in partial fraction expansion form,

$$Z(p) = \frac{k_0}{p} + \sum_{i=1}^{m} \frac{2k_i p}{(p^2 + \omega_i^2)} + k_\infty p \qquad (3.4)$$

We may now use the results of Section 2.7 to extract *all* the poles of $Z(p)$ in (3.4). After they have been extracted, we are left with a short-circuit (the remainder $Z_1(p) = 0$ in (2.57) and Fig. 2.8), and the synthesis of the whole function is completed. The network interpretation of Fig. 2.8 is valid in this case for the entire impedance. One obtains the well known Foster first form shown in Fig. 3.1, where the element values are given by

$$C_0 = 1/k_0, \qquad L_\infty = k_\infty$$
$$C_i = 1/2k_i, \qquad L_i = 2k_i/\omega_i^2 \qquad i = 1, 2, \ldots, m \qquad (3.5)$$

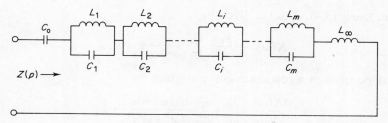

Fig. 3.1 Foster's first form.

where the ks are the residues in (3.4) which may be calculated using expressions (2.54) to (2.56).

Similarly, the above procedure can be applied to the admittance $Y(p) = 1/Z(p)$, which has a partial fraction expansion of the form,

$$Y(p) = \frac{h_0}{p} + \sum_{i=1}^{n} \frac{2h_i p}{(p^2 + \omega_i^2)} + h_\infty p \tag{3.6}$$

After the extraction of *all* the poles, we are left with an open-circuit ($Y_1 = 0$ in (2.61) and Fig. 2.10). The result is the Foster second form shown in Fig. 3.2 where the element values are given by

$$\begin{aligned} L_0 &= 1/h_0, \qquad C_\infty = h_\infty \\ L_i &= 1/2h_i, \qquad C_i = 2h_i/\omega_i^2 \end{aligned} \tag{3.7}$$

and the hs are the residues in (3.6) which may be calculated using the same expressions (2.54) to (2.56) but with $Z(p)$ replaced by $Y(p)$.

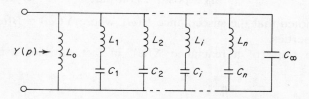

Fig. 3.2 Foster's second form.

In both Foster's realizations, the extraction of $j\omega$-axis poles reduces the degree of the function by the number of extracted poles. Therefore, the Foster forms are canonic realizations, i.e. they contain the minimum possible number of reactive components, this being equal to the degree of the function.

3.4 FOSTER'S REACTANCE THEOREM

Consider the behaviour of a reactance function on the $j\omega$-axis; write

$$Z(j\omega) = jX(\omega) \tag{3.8}$$

where, using (3.4), $X(\omega)$ is given by

$$X(\omega) = -\frac{k_0}{\omega} + \sum_{i=1}^{m} \frac{2k_i\omega}{(\omega_i^2 - \omega^2)} + k_\infty\omega \tag{3.9}$$

The slope of such a reactance curve is given by

$$\frac{\mathrm{d}X(\omega)}{\mathrm{d}\omega} = \frac{k_0}{\omega^2} + \sum_{i=1}^{m} \frac{2k_i(\omega_i^2 + \omega^2)}{(\omega_i^2 - \omega^2)^2} + k_\infty \tag{3.10}$$

But the residues k_0, k_i, k_∞ are all positive. Therefore we have

$$\frac{\mathrm{d}X(\omega)}{\mathrm{d}\omega} \geq 0 \tag{3.11}$$

where the equality can only be satisfied at $\omega = \infty$. Thus, the reactance $X(\omega)$ is a non-decreasing function, except at the poles. At the poles and zeros, the sign of $X(\omega)$ changes because the poles and zeros are simple. This is possible, while still satisfying (3.11), only if the poles and zeros alternate, i.e. a pole is followed by a zero, followed by a pole, etc. The general behaviour of $X(\omega)$ is shown in Fig. 3.3. Therefore we can write

$$Z(p) = H\left\{\frac{(p^2 + \omega_1^2)(p^2 + \omega_3^2) \cdots (p^2 + \omega_{2n-1}^2)}{p(p^2 + \omega_2^2)(p^2 + \omega_4^2) \cdots (p^2 + \omega_{2n-2}^2)}\right\}^{\pm 1} \tag{3.12a}$$

with

$$0 \leq \omega_1 < \omega_2 < \omega_3 < \omega_4 < \ldots \tag{3.12b}$$

and

$$H \text{ a positive constant} \tag{3.12c}$$

It is to be noted that the susceptance $B(\omega)$, where $Y(j\omega) = jB(\omega)$, possesses similar properties.

Now, the slope of a reactance $X(\omega)$, in fact, satisfies a more severe

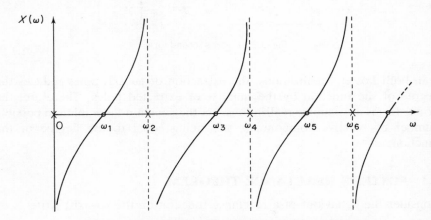

Fig. 3.3   General behaviour of a reactance (Foster) function.

requirement than that in (3.11). This is obtained by combining (3.9) and (3.10) to give

$$\frac{dX(\omega)}{d\omega} - \frac{X(\omega)}{\omega} = \frac{2k_0}{\omega} + \sum_{i=1}^{m} \frac{4k_i\omega^2}{(\omega_i^2 - \omega^2)^2} \tag{3.13}$$

and

$$\frac{dX(\omega)}{d(\omega)} + \frac{X(\omega)}{\omega} = 2k_\infty + \sum_{i=1}^{m} \frac{4k_i\omega_i^2}{(\omega_i^2 - \omega^2)^2} \tag{3.14}$$

Hence

$$\frac{dX(\omega)}{d\omega} \geq \frac{|X(\omega)|}{\omega}, \qquad \omega > 0 \tag{3.15}$$

But the residues of $Z(p)$ at its poles are given by[3,24]

$$k_i = \left\{ \frac{dY(p)}{dp} \right\}^{-1}_{p=j\omega_i} \tag{3.16}$$

and those of $Y(p)$ at its poles are

$$h_i = \left\{ \frac{dZ(p)}{dp} \right\}^{-1}_{p=j\omega_i} \tag{3.17}$$

Therefore, the positive slope property (3.11) guarantees that the residues are positive. But we have seen that (3.11) leads to the alternation property of the poles and zeros. It follows that this alternation property is equivalent to the statement that the residues of the function at its poles are positive. Thus, we have the following result.

Theorem 3.1 (Foster's Reactance Theorem)[27] *For the function $Z(p)$ to be realizable as the driving-point impedance of a lossless one-port, the necessary and sufficient condition is that $Z(p)$ be expressible as in (3.12), i.e.*

(i) $Z(p)$ *is real and rational*
(ii) $Z(p)$ *has only simple poles and zeros, alternating on the entire $j\omega$-axis (so they include $p = 0$ and $p = \infty$), and the scaling factor H is positive.*

3.5 CAUER'S CANONIC FORMS

The Foster realizations in Figs 3.1 and 3.2 are simply the network interpretation of the partial fraction expansion of the function. Equivalently, these are the result of successive extraction of the poles. We now consider another type of canonic realizations which employ the continued fraction expansion, or the successive removal of poles at $p = \infty$ (or $p = 0$) from the function and the subsequently inverted remainders.

Consider the case where $Z(p)$ has a pole at $p = \infty$. If it had a zero at $p = \infty$, its reciprocal has a pole at this point and the realization would proceed from the admittance. Thus $Z(p)$ is of the form

$$Z(p) = \frac{a_n p^n + a_{n-2} p^{n-2} + \ldots}{a_{n-1} p^{n-1} + a_{n-3} p^{n-3} + \ldots} \tag{3.18}$$

Performing a long division we have

$$Z(p) = \left(\frac{a_n}{a_{n-1}}\right) p + Z_1(p) \tag{3.19}$$

where

$$Z_1(p) = \frac{a'_{n-2} p^{n-2} + a'_{n-4} p^{n-4} + \ldots}{a_{n-1} p^{n-1} + a_{n-3} p^{n-3} + \ldots} \tag{3.20}$$

By Corollary 2.7, the remainder $Z_1(p)$ is also a reactance function (remember that a reactance function is a special case of a p.r.f.). Clearly, from (3.20), $Y_1(p) = 1/Z_1(p)$ has a pole at $p = \infty$, so that the process can be iterated leading to a continued fraction expansion of the form

$$Z(p) = b_1 p + \frac{1}{Y_1(p)}$$

$$= b_1 p + \cfrac{1}{b_2 p + \cfrac{1}{b_3 p + \cdots \cfrac{}{b_n p}}} \tag{3.21}$$

Noting that $b_1 p$ is the impedance of an inductor, $b_2 p$ is the admittance of a capacitor, etc. we are immediately led to the network interpretation of the expansion in (3.21) shown in Fig. 3.4. This is Cauer's first form. It is a ladder structure in which all the series branches are inductors and the shunt ones

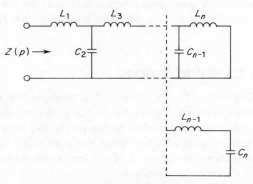

Fig. 3.4   Cauer's first form.

are capacitors. The element values are simply the coefficients of p in expansion (3.21).

Clearly, expansion (3.21) yields all b_i positive since the residue of a Foster function (impedance or admittance) at $p = \infty$ is guaranteed positive. Alternatively we know from Corollary 2.8 in Section 2.11 that the sum of numerator and denominator of a p.r.f. is a strictly Hurwitz polynomial. Therefore, according to the Hurwitz test in (2.50), the continued fraction expansion of $Z(p)$ in (3.21) must yield quotients satisfying (2.51), i.e. all are strictly positive. This interpretation coupled with Theorem 3.1 lead to the following result.

Corollary 3.1 *Let $Q(p)$ be a strictly Hurwitz polynomial and write*

$$Q(p) = m(p) + n(p)$$

where $m(p)$ is even and $n(p)$ is odd. Then

$$Z(p) = \frac{m(p)}{n(p)}$$

is a reactance function. Conversely, if $Z(p)$ is a reactance function, then the sum of its numerator and denominator is a strictly Hurwitz polynomial.

Next consider an alternative ladder development of a reactance function. If $Z(p)$ has a pole at $p = 0$, the continued fraction expansion is used to extract poles at $p = 0$ from $Z(p)$ and the subsequently inverted remainders. If $Z(p)$ has a zero at $p = 0$, then $Y(p) = 1/Z(p)$ has a pole at this point and the first step is an inversion. Arranging the numerator and denominator in ascending powers of p and performing the continued fraction expansion around $p = 0$ we obtain

$$Z(p) = \frac{a_0 + a_2 p^2 + a_4 p^4 + \ldots}{a_1 p + a_3 p^3 + a_5 p^5 + \ldots}$$

$$= \frac{1}{d_1 p} + Z_1(p)$$

$$= (1/d_1 p) + \frac{1}{Y_1(p)}$$

$$= (1/d_1 p) + \cfrac{1}{(1/d_2 p) + \cfrac{1}{\ddots \cfrac{}{(1/d_n p)}}} \tag{3.22}$$

The expansion in (3.22) gives Cauer's second form shown in Fig. 3.5. Again, this is a ladder structure, but the series branches are capacitors and the shunt ones are inductors.

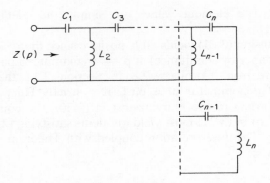

Fig. 3.5   Cauer's second form.

In each of the Cauer realizations, the number of elements is equal to the degree of the function. Therefore, like the Foster forms, these are also canonic realizations. It is also to be noted that mixed realizations can be obtained, which employ both the extraction of poles at $p = \infty$ (or $p = 0$) and finite $j\omega$-axis poles of $Z(p)$ (or $Y(p)$). These also lead to canonic forms.

Now, using the synthesis technique leading to the Cauer realizations, it is possible to derive an alternative, easily verifiable, set of conditions for an impedance to be realizable as a lossless one-port. Thus, from Theorem 3.1 and the Cauer realizations we have the following result.

Corollary 3.2   *A real rational function $Z(p)$ is realizable as the driving-point impedance of a lossless one-port, if and only if*

(a) Ev $Z(p) = 0$
(b) *The sum of the numerator and denominator of $Z(p)$ is a strictly Hurwitz polynomial.*

The necessity of the above conditions has already been established. Sufficiency follows directly from Corollary 3.1 and expansion (3.21) as demanded by Theorem 3.1. The conditions of Corollary 3.2 have the advantage over those of Theorem 3.1 in that the former avoid the task of factorizing the numerator and denominator of $Z(p)$.

Example 3.1   Verify that the following impedance is a reactance (Foster) function and find the Foster and Cauer realizations. Also realize the impedance in the form shown in Fig. 3.6.

$$Z(p) = \frac{p(p^2 + 2)(p^2 + 4)}{(p^2 + 1)(p^2 + 3)}$$

Solution   The given impedance has a positive multiplier ($= 1$) and its zeros and poles are simple, occurring in conjugate pairs (for Z to be real) and

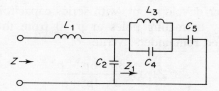

Fig. 3.6   Pertinent to Example 3.1.

alternating on the $j\omega$-axis with a zero at $p=0$ and a pole at $p=\infty$. Thus, $Z(p)$ satisfies the conditions of Theorem 3.1 and is a Foster function. To find the two Foster forms $Z(p)$ and $Y(p)$ are expanded in partial fraction form, and the residues are calculated using (2.54) to (2.56). Thus

$$Z(p) = \frac{\frac{3}{2}p}{p^2+1} + \frac{\frac{1}{2}p}{p^3+3} + p$$

$$Y(p) = \frac{\frac{3}{8}}{p} + \frac{\frac{1}{4}p}{p^2+2} + \frac{\frac{3}{8}p}{p^2+4}$$

Using the expressions for the element values in (3.5) and (3.7), the networks shown in Fig. 3.7(a), (b) are obtained.

We now obtain the Cauer realization with series inductors and shunt capacitors. Poles at $p=\infty$ are extracted from $Z(p)$ and the subsequently inverted remainders. The continued fraction expansion of $Z(p)$ around $p=\infty$ gives

$$
\begin{array}{r}
p \\
p^4+4p^2+3 \overline{\smash{\big)}\, p^5+6p^3+8p} \\
p^5+4p^3+3p \\
\hline
\end{array}
$$

$$
\begin{array}{r}
\frac{1}{2}p \\
2p^3+5p \overline{\smash{\big)}\, p^4+4p^2+3} \\
p^4+\frac{5}{2}p^2 \\
\hline
\end{array}
$$

$$
\begin{array}{r}
\frac{4}{3}p \\
\frac{3}{2}p^2+3 \overline{\smash{\big)}\, 2p^3+5p} \\
2p^3+4p \\
\hline
\end{array}
$$

$$
\begin{array}{r}
\frac{3}{2}p \\
p \overline{\smash{\big)}\, \frac{3}{2}p^2+3} \\
\frac{3}{2}p^2 \\
\hline
\end{array}
$$

$$
\begin{array}{r}
\frac{1}{3}p \\
3 \overline{\smash{\big)}\, p} \\
p \\
\hline
0
\end{array}
$$

The network, with the element values, is shown in Fig. 3.7(c).

The second ladder development, with series capacitors and shunt inductors, is obtained by extracting poles at $p = 0$ from the function and the subsequently inverted remainders. Write

$$Z(p) = \frac{8p + 6p^3 + p^5}{3 + 4p^2 + p^4}$$

Since $Z(p)$ has a zero at $p = 0$, $Y(p)$ has a pole at $p = 0$ and we begin by inverting $Z(p)$. Thus

$$Y(p) = \frac{3 + 4p^2 + p^4}{8p + 6p^3 + p^5}$$

The continued fraction expansion of $Y(p)$ around $p = 0$ is obtained as

$$
\begin{array}{r}
\dfrac{1/2.67p}{} \\
8p + 6p^3 + p^5 \; \overline{\smash{\big)}\; 3 + 4p^2 + p^4} \\
3 + 2.25p^2 + 0.375p^4 \\
\end{array}
$$

$$
\begin{array}{r}
\dfrac{1/0.218p}{} \\
1.75p^2 + 0.625p^4 \; \overline{\smash{\big)}\; 8p + 6p^3 + p^5} \\
8p + 2.86p^3 \\
\hline
3.14p^3 + p^5
\end{array}
$$

$$
\begin{array}{r}
\dfrac{1/1.794p}{} \\
3.14p^3 + p^5 \; \overline{\smash{\big)}\; 1.75p^2 + 0.625p^4} \\
1.75p^2 + 0.557p^4 \\
\end{array}
$$

$$
\begin{array}{r}
\dfrac{1/0.0217p}{} \\
0.068p^4 \; \overline{\smash{\big)}\; 3.143p^3 + p^5} \\
3.143p^3 \\
\end{array}
$$

$$
\begin{array}{r}
\dfrac{1/14.7p}{} \\
p^5 \; \overline{\smash{\big)}\; 0.068p^4} \\
0.068p^4 \\
\hline
0
\end{array}
$$

Remembering that the first term in the above expansion represents an admittance, we obtain the Cauer second form shown in Fig. 3.7(d).

Finally, we consider the realization of the same function in the required form in Fig. 3.6. First, extract a pole at $p = \infty$ from $Z(p)$, then a pole at $p = \infty$ from the inverted remainder. The element values are obtained by carrying out the first two steps in the continued fraction expansion of $Z(p)$ in exactly the same manner as in the part of this problem corresponding to Fig. 3.7(c). However, we terminate the process after two steps and evaluate the remain-

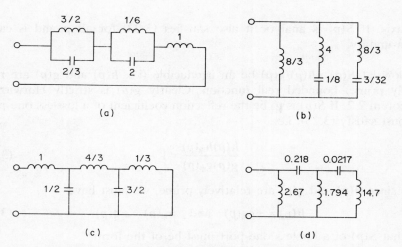

Fig. 3.7 Four canonic realizations of the impedance of Example 3.1 (element values in Henrys and Farads)

der. Thus, the first quotients are p and $\frac{1}{2}p$; the remainder is given by

$$Z_1(p) = \frac{3p^2+6}{4p^3+10p} = \frac{3p^2+6}{4p(p^2+5/2)}$$

whose partial fraction expansion is given by

$$Z_1(p) = \frac{\frac{3}{5}}{p} + \frac{\frac{3}{20}p}{p^2+\frac{5}{2}}$$

The complete network of Fig. 3.6 has the element values:

$$L_1 = 1 \text{ H}, \quad C_2 = \tfrac{1}{2} \text{ F}, \quad L_3 = \tfrac{3}{50} \text{ H}, \quad C_4 = \tfrac{20}{3} \text{ F}, \quad C_5 = \tfrac{5}{3} \text{ F}.$$

3.6 THE REFLECTION COEFFICIENT OF A LOSSLESS ONE-PORT

If $Z(p)$ is a reactance function, then its even-part is identically zero as given by (3.3). Using (2.74) in (3.3) we conclude that the reflection coefficient (relative to $r_g = 1\,\Omega$) of a lossless one-port satisfies

$$1 - S(p)\, S_*(p) = 0 \tag{3.23a}$$

or

$$SS_* = 1 \tag{3.23b}$$

In particular, on the $j\omega$-axis (Re $p = 0$) we have

$$|S(j\omega)| = 1 \tag{3.24}$$

Definition 3.2 A function satisfying (3.24) is said to be unitary on the

58

$j\omega$-axis. If $S(p)$ is analytic it also satisfies (3.23) for all p and is called para-unitary.

Now let $S(p) = h(p)/g(p)$ be an irreducible (i.e. $h(p)$ and $g(p)$ are relatively prime) bounded real function. Clearly $g(p)$ is strictly Hurwitz by Theorem 2.2. If $S(p)$ is to be the reflection coefficient of a lossless one-port, it must satisfy (3.23), i.e.

$$\frac{h(p)h_*(p)}{g(p)g_*(p)} = 1 \tag{3.25}$$

But since $h(p)$ and $g(p)$ are relatively prime, we must have

$$h(p) = \pm g_*(p) \quad \text{and} \quad h_*(p) = \pm g(p) \tag{3.26}$$

so that $S(p)$ of a lossless one-port must be of the form

$$S(p) = \varepsilon \frac{g_*(p)}{g(p)}, \quad \varepsilon = \pm 1 \tag{3.27}$$

PROBLEMS

3.1 Show that each of the following functions is a reactance function, and find the Foster and Cauer forms.

(a) $$Z(p) = \frac{4p^4 + 13p^2 + 3}{p^5 + 5p^3 + 4p}$$

(b) $$Z(p) = \frac{(p^2+1)(p^2+3)}{p(p^2+2)(p^2+4)}$$

3.2 Two impedances $Z_1(p)$ and $Z_2(p)$ are said to be 'reciprocals' if they satisfy

$$Z_1(p)Z_2(p) = 1$$

If Z_1 is a Foster function given by

$$Z_1(p) = \frac{p(p^2+3)(p^2+5)}{(p^2+2)(p^2+4)}$$

realize $Z_1(p)$ and from the resulting network, construct a lossless one-port which realizes $Z_2(p)$.

3.3 Consider the Foster impedance

$$Z(p) = \frac{(p^2+1)(p^2+3)(p^2+6)}{p(2p^2+3)(p^2+5)}$$

Extract two successive poles at $p = \infty$ from Z and the inverted remainder. Evaluate the remaining impedance and realize it in Foster's first form.

3.4 A Foster impedance has poles at $\omega = 1000$ and $3000 \, \text{rad.s}^{-1}$. It also has a zero at $\omega = 2000 \, \text{rad.s}^{-1}$ at which the slope of the function is $0.6 \, \Omega \, \text{rad}^{-1}.\text{s}$. Find two canonic realizations of the function.

3.5 Derive the necessary and sufficient conditions under which a given impedance is realizable as the driving-point impedance of an RC one-port, i.e. containing resistors and capacitors only. In particular, show that the poles and zeros of such an impedance must alternate on the negative real axis, with the lowest critical frequency being a pole.

Chapter 4

Fundamental Properties of Passive Two-ports

4.1 INTRODUCTION

This chapter is concerned with the fundamental properties of the matrices which may be used to characterize passive two-port networks (abbreviated two-ports). Particular emphasis is laid on the scattering description. The study in the present chapter is developed along similar lines to those in Chapter 2 for passive one-ports. The discussion represents significant generalizations and extension of the positive real and bounded real concepts to the matrices which describe the external behaviour of passive two-ports.

4.2 THE IMPEDANCE AND ADMITTANCE MATRICES

Consider the two-port N shown in Fig. 4.1, with the voltage and current variables given by the column matrices

$$[V(p)] = \begin{bmatrix} V_1(p) \\ V_2(p) \end{bmatrix} \tag{4.1}$$

$$[I(p)] = \begin{bmatrix} I_1(p) \\ I_2(p) \end{bmatrix} \tag{4.2}$$

Again, we assume that the sources can exist only at the ports, while the two-port N is passive. From the results of network analysis,[20] the set of equations relating the port variables may be written as

$$[D(p)][V(p)] = [E(p)][I(p)] \tag{4.3}$$

where $[D(p)]$ and $[E(p)]$ are square matrices of order 2, whose entries are *real polynomials*.

Now, if $\det[D(p)]$ does not vanish identically, it is possible to write

$$[V(p)] = [D(p)]^{-1}[E(p)][I(p)] \tag{4.4}$$

or

$$[V(p)] = [Z(p)][I(p)] \tag{4.5}$$

where $[Z(p)]$ is the familiar impedance matrix of the two-port, discussed

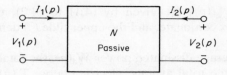

Fig. 4.1   A passive two-port.

briefly in Chapter 1, i.e.

$$[Z(p)] = \begin{bmatrix} z_{11}(p) & z_{12}(p) \\ z_{21}(p) & z_{22}(p) \end{bmatrix} \tag{4.6}$$

with its entries (the z-parameters) being *rational functions* for a passive two-port.

Similarly, if in (4.3) det $[E(p)]$ does not vanish identically, we may write

$$[I(p)] = [E(p)]^{-1}[D(p)][V(p)] \tag{4.7}$$

or

$$[I(p)] = [Y(p)][V(p)] \tag{4.8}$$

where $[Y(p)]$ is the admittance matrix of the two-port,

$$[Y(p)] = \begin{bmatrix} y_{11}(p) & y_{12}(p) \\ y_{21}(p) & y_{22}(p) \end{bmatrix} \tag{4.9}$$

and its entries (the y-parameters) are *rational functions* for a passive two-port.

In certain situations, det $[D(p)]$ in (4.4) vanishes identically, and the network does not possess an impedance matrix. Similarly det $[E(p)]$ in (4.7) may vanish identically and the network does not possess an admittance representation. It is also possible for a network to possess neither an impedance nor an admittance matrix. One such network is the ideal transformer. However, *a useful property of the matrices $[D(p)]$ and $[E(p)]$ is that* det $\{[D(p)]+[E(p)]\}$ *cannot vanish identically.*

Clearly from (4.5) and (4.8), the impedance and admittance matrices are related by

$$[Y(p)] = [Z(p)]^{-1}, \quad [Z(p)] = [Y(p)]^{-1} \tag{4.10}$$

It is also worth remembering that if the two-port N is *reciprocal*, then

$$z_{12} = z_{21}, \quad y_{12} = y_{21} \tag{4.11}$$

i.e. both $[Z(p)]$ and $[Y(p)]$ are symmetric.

We now proceed along similar lines to those followed for one-ports in Chapter 2, to derive the fundamental properties of the impedance and admittance matrices. The total complex power entering the network N through both ports is given by

$$W(p) = V_1(p)I_1^*(p) + V_2(p)I_2^*(p)$$
$$= [\tilde{I}(p)][V(p)] \tag{4.12}$$

where $[V(p)]$ and $[I(p)]$ are given by (4.1) to (4.2), the upper asterisk denotes the complex conjugate, and the upper tilde ($\tilde{\ }$) denotes the conjugate transpose.

Now the total average dissipated power $W_d(p)$, the total average magnetic enegy $T_m(p)$, and the total average electric energy $T_e(p)$ summed over all elements of N were obtained in expressions (2.19) to (2.31), and are all real and positive quantities for all p. Moreover, the two-port together with the sources conserve complex power.[20] Hence,

$$[\tilde{I}][V] = W_d + 2pT_m + 2p^*T_e \tag{4.13}$$

where the argument p has been dropped for convenience. This practice is adopted henceforth, wherever no confusion may arise.

If the two-port possesses an impedance matrix $[Z]$, then (4.5) can be used with (4.13) to give

$$[\tilde{I}][Z][I] = W_d + 2\sigma(T_m + T_e) + 2j\omega(T_m - T_e)$$
$$= F(p) \tag{4.14}$$

Noting that the quantities W_d, T_m, and T_e are real and positive for all p, then taking the conjugate transpose of (4.14) we have

$$[\tilde{I}][\tilde{Z}][I] = W_d + 2\sigma(T_m + T_e) - 2j\omega(T_m - T_e) \tag{4.15}$$

Adding (4.14) and (4.15) we obtain

$$[\tilde{I}]\{[Z] + [\tilde{Z}]\}[I] = 2W_d + 4\sigma(T_m + T_e) \tag{4.16}$$

But the right-hand side of (4.16) is positive for all positive values of σ ($\sigma = \operatorname{Re} p$). Therefore the Hermitian matrix

$$[G(p)] = [Z(p)] + [\tilde{Z}(p)] \tag{4.17}$$

is positive definite in $\operatorname{Re} p > 0$ (see Appendix A.1 for a brief discussion of Hermitian forms). The similarity to the one-port case is now evident, and this leads to the following definition.

Definition 4.1 A real matrix $[Z(p)]$ (i.e. real for p real) such that $[Z(p)] + [\tilde{Z}(p)]$ is positive definite in $\operatorname{Re} p > 0$, is called a positive real matrix.

In a similar manner, we can show that if the two-port possesses an admittance matrix $[Y(p)]$ then

$$[\tilde{V}]\{[Y] + [\tilde{Y}]\}[V] = 2W_d + 4\sigma(T_m + T_e) \tag{4.18}$$

from which $[Y]$ is also a positive real matrix.

Theorem 4.1 *The impedance and admittance matrices of a passive, lumped, linear, and time-invariant two-port, are both positive real matrices.*

Definition 4.2 A matrix $[Z]$ whose entries are rational functions of p is said to have a pole at a point p_0 if *some* entry z_{ij} has a pole at p_0.

Definition 4.3 A matrix $[Z]$ is said to be analytic in a region of the p-plane if *all* its entries are analytic functions in that region.

With the above definitions in mind, we observe from (4.14) that the poles of $[Z]$ correspond to poles of $F(p)$ and vice versa. It also follows that, since $F(p)$ is a positive real function, $[Z]$ is analytic in $\text{Re}\, p > 0$ and it may have simple poles on the $j\omega$-axis. The residue of $F(p)$ at a $j\omega$-axis pole is given by

$$
\lim_{p \to j\omega_i} (p - j\omega_i)[I_1^* \quad I_2^*]\begin{bmatrix} z_{11} & z_{12} \\ z_{21} & z_{22} \end{bmatrix}\begin{bmatrix} I_1 \\ I_2 \end{bmatrix}
$$

$$
= [I_1^* \quad I_2^*]\begin{bmatrix} k_{11} & k_{12} \\ k_{21} & k_{22} \end{bmatrix}\begin{bmatrix} I_1 \\ I_2 \end{bmatrix}
$$

$$
= [\tilde{I}][K][I] \tag{4.19}
$$

where $[K]$ is the matrix of the residues of the entries of $[Z]$. The residue of $F(p)$ is, therefore, a quadratic form $[\tilde{I}][K][I]$. This must be real and non-negative for any $[I]$. Hence, the residue matrix $[K]$ must be positive semi-definite (see Appendix A.1).

4.3 THE SCATTERING MATRIX

Although the impedance and admittance matrices lead to useful and interesting conclusions, the characterization of two-ports in terms of these matrices suffers from a number of disadvantages. First, the existence of such matrices is not guaranteed for all networks. Secondly, in filter design the required two-port is a doubly terminated one, i.e. it is driven by a source of known internal resistance and terminated in a resistive load. In this case, the parameter of interest is a transfer function in terms of which conditions regarding the performance of the network can be conveniently formulated. For these reasons, we now introduce a matrix description of two-ports, which has been shown to exist for any passive two-port.[19,20] This is the *scattering matrix*, whose entries are the *scattering parameters*. They are particularly useful in the description of power transfer under practical terminating conditions, therefore they are exclusively used in filter synthesis.

Consider again the two-port shown in Fig. 4.1 where the port currents and voltages are given by (4.1) to (4.2). Let these variables be normalized with respect to the *strictly positive* resistances, r_1 and r_2 as follows

$$
V_{n1} = \frac{V_1}{\sqrt{r_1}}, \qquad V_{n2} = \frac{V_2}{\sqrt{r_2}} \tag{4.20}
$$

$$
I_{n1} = \sqrt{r_1}\, I_1, \qquad I_{n2} = \sqrt{r_2}\, I_2 \tag{4.21}
$$

Thus, we can write the normalized column matrices as

$$
[V_n] = [r^{1/2}]^{-1}[V] \tag{4.22}
$$

$$
[I_n] = [r^{1/2}][I] \tag{4.23}
$$

where

$$[r^{1/2}] = \begin{bmatrix} r_1^{1/2} & 0 \\ 0 & r_2^{1/2} \end{bmatrix} \tag{4.24}$$

We had our first experience of this type of normalization in Section 2.10 where the reflection coefficient of a one-port was introduced. Using relations similar to (2.79) we define the incident signals at the ports by

$$[\alpha] = [V_n] + [I_n] \tag{4.25a}$$

where

$$[\alpha] = \begin{bmatrix} \alpha_1 \\ \alpha_2 \end{bmatrix} \tag{4.25b}$$

Also the reflected signals at the ports are defined by

$$[\beta] = [V_n] - [I_n] \tag{4.26a}$$

where

$$[\beta] = \begin{bmatrix} \beta_1 \\ \beta_2 \end{bmatrix} \tag{4.26b}$$

From (4.25) and (4.26) we may write

$$[V_n] = \tfrac{1}{2}\{[\alpha] + [\beta]\} \tag{4.27}$$

$$[I_n] = \tfrac{1}{2}\{[\alpha] - [\beta]\} \tag{4.28}$$

Now, relating $[V_n]$ and $[I_n]$ using two polynomial matrices by a relation similar to (4.3) we have

$$[D_n][V_n] = [E_n][I_n] \tag{4.29}$$

which upon use of (4.27) to (4.28) gives

$$\{[E_n] - [D_n]\}[\alpha] = \{[E_n] + [D_n]\}[\beta] \tag{4.30}$$

and a classical result in network analysis[20] states that $\det\{[D_n] + [E_n]\}$ cannot vanish identically. Therefore we can write from (4.30)

$$[\beta(p)] = [S(p)][\alpha(p)] \tag{4.31}$$

with

$$[S(p)] = \{[E_n] + [D_n]\}^{-1}\{[E_n] - [D_n]\} \tag{4.32}$$

where $[S(p)]$ is the *scattering matrix* of the two-port

$$[S(p)] = \begin{bmatrix} S_{11}(p) & S_{12}(p) \\ S_{21}(p) & S_{22}(p) \end{bmatrix} \tag{4.33}$$

We now consider the physical interpretation of the entries of the scattering matrix. Write (4.31) explicitly as

$$\begin{aligned} \beta_1 &= S_{11}\alpha_1 + S_{12}\alpha_2 \\ \beta_2 &= S_{21}\alpha_1 + S_{22}\alpha_2 \end{aligned} \tag{4.34}$$

From the above expressions the scattering parameters are defined as

$$S_{11} = \frac{\beta_1}{\alpha_1}\bigg|_{\alpha_2=0}, \quad S_{12} = \frac{\beta_1}{\alpha_2}\bigg|_{\alpha_1=0}$$

$$S_{21} = \frac{\beta_2}{\alpha_1}\bigg|_{\alpha_2=0}, \quad S_{22} = \frac{\beta_2}{\alpha_2}\bigg|_{\alpha_1=0}$$

(4.35)

Hence, each parameter is defined as the ratio of a reflected signal to an incident signal under the condition of zero incident signal at the other port. To give a physical meaning to these parameters, we must find out what the condition of zero incident signal at a port means. Letting $\alpha_2 = 0$ in (4.25) gives $V_{n2} = -I_{n2}$ or $r_2^{-1/2}V_2 = -r_2^{1/2}I_2$, i.e. $V_2 = -r_2 I_2$. Thus $\alpha_2 = 0$ means that the output port is terminated in a resistor of value equal to the reference normalizing resistor r_2. Similarly $\alpha_1 = 0$ means that the input port is terminated in its reference resistor r_1.

Specifically, from (4.35), $S_{11}(p)$ is the ratio of reflected to incident signals at the input port, when the output port is match-terminated. Therefore $S_{11}(p)$ is the *reflection coefficient* of the one-port formed by terminating the two-port in its reference resistor r_2. Similarly $S_{22}(p)$ is the *reflection coefficient* of the one-port produced by closing the input port on its reference resistor r_1. This network interpretation is shown in Fig. 4.2. Thus $S_{11}(p)$ is the *input reflection coefficient* and $S_{22}(p)$ is the *output reflection coefficient*.

Now consider the situation of Fig. 4.3 in which the two-port is driven by a source of internal resistance r_g, and terminated at the output in a resistive load r_ℓ. If we choose the port normalizing numbers to be equal to the actual termination, we have $r_1 = r_g$ and $r_2 = r_\ell$. Thus, in Fig. 4.3

$$V_2 = -r_\ell I_2 \tag{4.36}$$

and from (4.22), (4.23), and (4.25) we have

$$\alpha_1 = r_g^{-1/2}(V_1 + r_g I_1) \tag{4.37}$$

$$\alpha_2 = r_\ell^{-1/2}(V_2 + r_\ell I_2) \tag{4.38}$$

and use of (4.36) in (4.38) gives

$$\alpha_2 = 0 \tag{4.39}$$

Combining (4.27) and (4.39) we obtain

$$V_{n2} = \tfrac{1}{2}\beta_2 \tag{4.40}$$

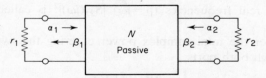

Fig. 4.2 Pertinent to the interpretation of S_{11} and S_{22}

66

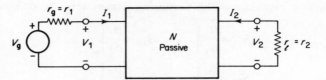

Fig. 4.3   Pertinent to the interpretation of S_{21}

or, in terms of the denormalized voltage V_2,

$$V_2 = \frac{r_\ell^{-1/2}}{2} \beta_2 \tag{4.41}$$

Also the input conditions in Fig. 4.3 satisfy

$$V_g = V_1 + r_g I_1 \tag{4.42}$$

which, using (4.37), becomes

$$V_g = r_g^{1/2} \alpha_1 \tag{4.43}$$

Dividing (4.41) by (4.43) we have, under the condition of (4.39)

$$\left.\frac{\beta_2}{\alpha_1}\right|_{\alpha_2=0} = 2\sqrt{\frac{r_g}{r_\ell}} \frac{V_2}{V_g} \tag{4.44}$$

Therefore, the parameter $S_{21}(p)$ is given by

$$S_{21}(p) = 2\sqrt{\frac{r_g}{r_\ell}} \frac{V_2(p)}{V_g(p)} \tag{4.45}$$

and the square of its modulus at $p = j\omega$,

$$|S_{21}(j\omega)|^2 = \frac{4r_g}{r_\ell} \frac{|V_2(j\omega)|^2}{|V_g(j\omega)|^2}$$

$$= \frac{|V_2(j\omega)|^2/r_\ell}{|V_g(j\omega)|^2/4r_g} \tag{4.46}$$

is the ratio of the power delivered to the load to the maximum power available from the source at the input port, under reference terminating conditions.

Similarly, if we interchange the ports 1 and 2 we arrive at a correponding interpretation of $S_{12}(p)$. Due to these interpretations, $S_{21}(p)$ is called the *forward transmission coefficient,* and $S_{12}(p)$ is called the *reverse transmission coefficient.* At real frequencies $(p = j\omega)$ $|S_{21}(j\omega)|^2$ is called the *transducer power gain.*

Next consider the total complex power entering the two-port shown in Fig. 4.1 through both ports,

$$W = [\tilde{I}][V]$$

$$= [\tilde{I}_n][V_n] \tag{4.47}$$

Substituting for $[V_n]$ and $[I_n]$ from (4.27) and (4.28) into (4.47) we obtain

$$4W = [\tilde{\alpha}][\alpha] - [\tilde{\beta}][\beta] - [\tilde{\beta}][\alpha] + [\tilde{\alpha}][\beta] \qquad (4.48)$$

Hence, the real part of the complex power is given by

$$\mathrm{Re}\,W = \tfrac{1}{4}\{[\tilde{\alpha}][\alpha] - [\tilde{\beta}][\beta]\} \qquad (4.49)$$

and using (4.31) in (4.49) we obtain

$$\mathrm{Re}\,W = \tfrac{1}{4}\{[\tilde{\alpha}][\alpha] - [\tilde{\alpha}][\tilde{S}][S][\alpha]\}$$
$$= \tfrac{1}{4}[\tilde{\alpha}]\{[U_2] - [\tilde{S}][S]\}[\alpha] \qquad (4.50)$$

where $[U_2]$ is the 2×2 unit matrix. But we have seen that for a passive two-port, $\mathrm{Re}\,W$ is positive for values of p with positive real part. It follows that the Hermitian matrix (see Appendix A.1),

$$[Q(p)] = [U_2] - [\tilde{S}(p)][S(p)] \qquad (4.51)$$

is positive definite in $\mathrm{Re}\,p > 0$.

Definition 4.4 A real matrix $[S(p)]$ such that $[U] - [\tilde{S}][S]$ is positive definite in $\mathrm{Re}\,p > 0$ is called a bounded real matrix.

According to the above definition and the preceding analysis, we can assert that the scattering matrix $[S(p)]$ of a passive two-port, referred to real terminations, is a *bounded real* matrix.

Now the diagonal entries of the matrix $[Q(p)]$ in (4.51) are given by

$$\begin{aligned} q_{11} &= 1 - (|S_{11}|^2 + |S_{21}|^2) \\ q_{22} &= 1 - (|S_{12}|^2 + |S_{22}|^2) \end{aligned} \qquad (4.52)$$

But since $[Q]$ is positive definite in $\mathrm{Re}\,p > 0$, then its principal minors must be positive in $\mathrm{Re}\,p > 0$. Therefore the diagonal entries q_{ii} ($i = 1, 2$) must be positive in $\mathrm{Re}\,p > 0$. Thus, from (4.52)

$$\begin{aligned} |S_{11}|^2 + |S_{21}|^2 &< 1 \qquad \mathrm{Re}\,p > 0 \\ |S_{12}|^2 + |S_{22}|^2 &< 1 \qquad \mathrm{Re}\,p > 0 \end{aligned} \qquad (4.53)$$

so that

$$|S_{ij}(p)|^2 < 1 \qquad \mathrm{Re}\,p > 0 \qquad (4.54)$$

This means that every entry of the scattering matrix of a passive two-port, is a bounded real function in the meaning of Definition 2.4. In particular on the imaginary axis

$$|S_{ij}(j\omega)| \leq 1 \qquad -\infty \leq \omega \leq \infty \qquad (4.55)$$

and as demanded by Theorem 2.2, every entry of $[S(p)]$ is analytic in $\mathrm{Re}\,p \geq 0$, i.e. in the closed right half-plane including the point at ∞.

4.4 RELATIONSHIPS BETWEEN $[S]$, $[Z]$, AND $[Y]$

Consider the relationship between the scattering matrix of a two-port and its impedance matrix. From (4.22), (4.23), (4.27), and (4.28) we have

$$2[V] = [r^{1/2}]\{[\alpha] + [\beta]\}$$
$$2[I] = [r^{1/2}]^{-1}\{[\alpha] - [\beta]\} \tag{4.56}$$

Substituting for $[V]$ and $[I]$ from the above expressions into the impedance equation (4.5) we obtain

$$[r^{1/2}]\{[\alpha] + [\beta]\} = [Z][r^{1/2}]^{-1}\{[\alpha] - [\beta]\} \tag{4.57}$$

and use of (4.31) in (4.57) gives

$$\{[U_2] + [S]\}[\alpha] = [r^{1/2}]^{-1}[Z][r^{1/2}]^{-1}\{[U_2] - [S]\}[\alpha] \tag{4.58}$$

Therefore

$$[r^{1/2}]^{-1}[Z][r^{1/2}]^{-1} = \{[U_2] + [S]\}\{[U_2] - [S]\}^{-1} \tag{4.59}$$

Defining $[\bar{Z}]$ as the *normalized impedance* matrix given by

$$[\bar{Z}] = [r^{1/2}]^{-1}[Z][r^{1/2}]^{-1} \tag{4.60}$$

its entries (taking $r_1 = r_g$, $r_2 = r_\ell$) are related to those of $[Z]$ by

$$\bar{z}_{11} = z_{11}/r_g, \qquad \bar{z}_{12} = z_{12}/\sqrt{r_g r_\ell}$$
$$\bar{z}_{21} = z_{21}/\sqrt{r_g r_\ell}, \qquad \bar{z}_{22} = z_{22}/r_\ell \tag{4.61}$$

Therefore (4.59) becomes

$$[\bar{Z}] = \{[U_2] + [S]\}\{[U_2] - [S]\}^{-1}$$
$$= 2\{[U_2] - [S]\}^{-1} - [U_2] \tag{4.62}$$

Writing the above expression explicitly we have

$$\bar{z}_{11} = \{(1 - S_{22})(1 + S_{11}) + S_{12}S_{21}\}/\Delta \tag{4.63a}$$

$$\bar{z}_{12} = 2S_{12}/\Delta, \quad \bar{z}_{21} = 2S_{21}/\Delta \tag{4.63b}$$

$$\bar{z}_{22} = \{(1 - S_{11})(1 + S_{22}) + S_{12}S_{21}\}/\Delta \tag{4.63c}$$

where

$$\Delta = (1 - S_{11})(1 - S_{22}) - S_{12}S_{21} \tag{4.63d}$$

In a similar manner, the admittance matrix $[Y]$ can be expressed in terms of the scattering matrix. Define the normalized admittance matrix $[\bar{Y}]$ by

$$[\bar{Y}] = [r^{1/2}][Y][r^{1/2}] \tag{4.64}$$

then using (4.10) with (4.62) we have

$$[\bar{Y}] = [\bar{Z}]^{-1}$$
$$= \{[U_2] - [S]\}\{[U_2] + [S]\}^{-1} \tag{4.65}$$

The inverse relations can be easily obtained as

$$[S] = \{[\bar{Z}] + [U_2]\}^{-1}\{[\bar{Z}] - [U_2]\}$$
$$= [U_2] - 2\{[\bar{Z}] + [U_2]\}^{-1} \tag{4.66}$$

and

$$[S] = \{[U_2] + [\bar{Y}]\}^{-1}\{[U_2] - [\bar{Y}]\}$$
$$= 2\{[U_2] + [\bar{Y}]\}^{-1} - [U_2] \tag{4.67}$$

Explicitly, (4.66) reads

$$S_{11} = \{(\bar{z}_{11} - 1)(\bar{z}_{22} + 1) - \bar{z}_{12}\bar{z}_{21}\}/\delta$$
$$S_{12} = 2\bar{z}_{12}/\delta, \quad S_{21} = 2\bar{z}_{21}/\delta \tag{4.68}$$
$$S_{22} = \{(\bar{z}_{11} + 1)(\bar{z}_{22} - 1) - \bar{z}_{12}\bar{z}_{21}\}/\delta$$

where

$$\delta = (\bar{z}_{11} + 1)(\bar{z}_{22} + 1) - \bar{z}_{12}\bar{z}_{21}$$

with similar expressions for (4.67).

Finally, for a *reciprocal* two-port use of (4.11) in (4.68) gives

$$S_{12}(p) = S_{21}(p) \tag{4.69}$$

i.e. the scattering matrix of a reciprocal two-port is *symmetric*.

4.5 THE TRANSMISSION MATRIX

An alternative method of relating the port variables of the two-port in Fig. 4.1 is to write

$$\begin{bmatrix} V_1 \\ I_1 \end{bmatrix} = [T(p)]\begin{bmatrix} V_2 \\ -I_2 \end{bmatrix} \tag{4.70}$$

where

$$[T(p)] = \begin{bmatrix} A(p) & B(p) \\ C(p) & D(p) \end{bmatrix} \tag{4.71}$$

is the *transmission* (or chain) matrix of the two-port. The reader must be familiar with the relationships between $[T]$, $[Z]$, and $[Y]$ from a course in network analysis. These are given below for later reference

$$\begin{bmatrix} A & B \\ C & D \end{bmatrix} = \frac{1}{z_{21}}\begin{bmatrix} z_{11} & \det[Z] \\ 1 & z_{22} \end{bmatrix} \tag{4.72}$$

$$= \frac{-1}{y_{21}}\begin{bmatrix} y_{22} & 1 \\ \det[y] & y_{11} \end{bmatrix} \tag{4.73}$$

$$\begin{bmatrix} z_{11} & z_{12} \\ z_{21} & z_{22} \end{bmatrix} = \frac{1}{C}\begin{bmatrix} A & \det[T] \\ 1 & D \end{bmatrix} \tag{4.74}$$

$$\begin{bmatrix} y_{11} & y_{12} \\ y_{21} & y_{22} \end{bmatrix} = \frac{1}{B}\begin{bmatrix} D & -\det[T] \\ -1 & A \end{bmatrix} \tag{4.75}$$

It is also possible to relate the transmission (chain) parameters to the scattering parameters. First, define the *normalized transmission matrix* $[\bar{T}(p)]$ as

$$[\bar{T}(p)] = \begin{bmatrix} \dfrac{1}{\sqrt{r_g}} & 0 \\ 0 & \sqrt{r_g} \end{bmatrix} [T(p)] \begin{bmatrix} \sqrt{r_\ell} & 0 \\ 0 & \dfrac{1}{\sqrt{r_\ell}} \end{bmatrix} \tag{4.76}$$

or

$$\begin{bmatrix} \bar{A} & \bar{B} \\ \bar{C} & \bar{D} \end{bmatrix} = \begin{bmatrix} \sqrt{\dfrac{r_\ell}{r_g}}A & \dfrac{B}{\sqrt{r_g r_\ell}} \\ \sqrt{r_g r_\ell}C & \sqrt{\dfrac{r_g}{r_\ell}}D \end{bmatrix} \tag{4.77}$$

Comparing (4.77) with (4.72) and using (4.63) we obtain for the normalized transmission parameters,

$$\bar{A} = \bar{z}_{11}/\bar{z}_{21} = \{(1 - S_{22})(1 + S_{11}) + S_{12}S_{21}\}/2S_{21} \tag{4.78a}$$

$$\bar{B} = \det[\bar{Z}]/\bar{z}_{21} = \{(1 + S_{11})(1 + S_{22}) - S_{12}S_{21}\}/2S_{21} \tag{4.78b}$$

$$\bar{C} = 1/\bar{z}_{21} = \Delta/2S_{21} \tag{4.78c}$$

$$\bar{D} = \bar{z}_{22}/\bar{z}_{21} = \{(1 - S_{11})(1 + S_{22}) + S_{12}S_{21}\}/2S_{21} \tag{4.78d}$$

with Δ given by (4.63d).

From (1.21), (4.72) to (4.73) and the rationality of the impedance or admittance matrix, we conclude that the transmission parameters of a passive two-port are all rational transfer functions. We also note from (4.74) that reciprocity ($z_{12} = z_{21}$) requires

$$\det[T] = 1 \tag{4.79}$$

The description of a two-port in terms of its transmission matrix is convenient when several networks are connected in cascade, since the overall transmission matrix is equal to the product of the transmission matrices of the individual two-ports. $[T]$ is also useful for the description of the ideal transformer which does not possess either an impedance or admittance representation.

PROBLEMS

4.1 Find the scattering matrix, referred to unit resistors for each of the networks shown in Fig. P4.1.

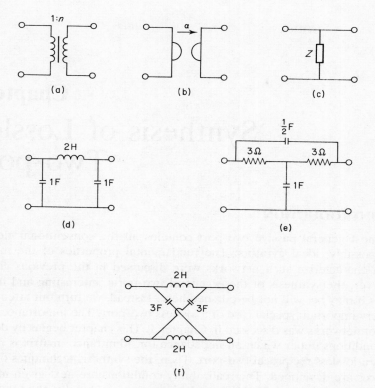

Fig. P4.1   Problem 4.1.

4.2 Consider a passive three-port as shown in Fig. P4.2. Extend the results of the scattering description in the present chapter, to show that the three-port is describable by a bounded real 3×3 scattering matrix.

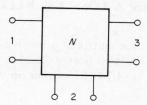

Fig. P4.2   Problem 4.2.

Chapter 5
Synthesis of Lossless Two-ports

5.1 INTRODUCTION

The most general passive two-port contains all the conventional elements and, possibly, ideal gyrators. The fundamental properties of the matrices which characterize such networks were discussed in the previous chapter. However, the synthesis of these general two-ports, interesting and instructive as it may be, will not be pursued here. Instead, we turn our attention to the very important special case of a lossless two-port. The importance of this class of networks was discussed in Chapter 1. This chapter begins by deriving the conditions under which an impedance (or admittance) matrix is realizable as a lossless reciprocal two-port. Then, the synthesis techniques of these matrices are developed. The realizability conditions are next given in terms of the scattering matrix and the transmission matrix. Finally, the fundamental properties and characterization of the doubly terminated lossless two-port are discussed together with their significance to the filter design problem. Related concepts such as the *zeros of transmission* and the *even-part function* are also discussed in detail.

5.2 THE IMPEDANCE MATRIX OF A LOSSLESS RECIPROCAL TWO-PORT

According to Theorem 4.1, the impedance matrix of a passive two-port is a positive real matrix. If, in addition, the two-port is lossless, there is no dissipation. Therefore, setting $W_d = 0$ in (4.14) we obtain on the imaginary axis ($\sigma = 0$),

$$F(j\omega) + F(-j\omega) = 0 \tag{5.1}$$

Therefore, by analytic continuation, we may write

$$F(p) + F_*(p) = 0 \tag{5.2}$$

where, as usual, the lower asterisk denotes replacing p by $-p$. Thus $F(p)$ is a reactance (Foster) function; therefore by Theorem 3.1 $F(p)$ has only simple poles on the $j\omega$-axis. Thus, from (4.14) the impedance matrix of the lossless

72

two-port has *all its poles* on the $j\omega$-axis and the residue matrix at each pole as given in (4.19), is positive semi-definite. Furthermore, from (4.16) with $W_d = 0$ we have

$$[Z(j\omega)] + [Z(-j\omega)]' = [0] \tag{5.3}$$

where the prime denotes the transpose. But since $[Z(p)]$ is analytic, we can apply analytic continuation to the entries of $[Z(p)]$ to yield

$$[Z(p)] + [\underset{\sim}{Z}(p)] = [0] \tag{5.4}$$

where the lower tilde denotes the transpose with p replaced by $-p$.

Expression (5.4) is true for any passive lossless two-port, *not necessarily* reciprocal. However, we are mainly interested in the reciprocal case. From (4.11) reciprocity requires that $[Z(p)]$ be symmetric. Therefore, $[Z] = [Z]'$ and (5.4) for a lossless reciprocal two-port becomes

$$[Z(p)] + [Z(p)]_* = [0] \tag{5.5}$$

The results of the above discussion are now put together in a theorem.

Theorem 5.1 *The 2×2 impedance matrix $[Z(p)]$ of a passive lumped lossless reciprocal two-port is real, symmetric, and rational expressible in the form*

$$[Z(p)] = \frac{[K_0]}{p} + 2p \sum_{i=1}^{m} \frac{[K_i]}{p^2 + \omega_i^2} + [K_\infty]p \tag{5.6}$$

where the residue matrices $[K_0]$, $[K_i]$, and $[K_\infty]$ are all real constant symmetric positive semi-definite matrices. Therefore, each residue matrix obtained according to (4.19),

$$[K] = \begin{bmatrix} k_{11} & k_{12} \\ k_{12} & k_{22} \end{bmatrix} \tag{5.7}$$

satisfies

$$k_{11} \geq 0, \quad k_{22} \geq 0 \tag{5.8a}$$

$$k_{11}k_{22} - k_{12}^2 \geq 0 \tag{5.8b}$$

in which k_{12} may be negative. At a pole which is common to all entries of $[Z]$, $k_{11} > 0$, $k_{22} > 0$, and (5.8b) is still in force.

Definition 5.1 By analogy with the impedance of a lossless one-port, a matrix satisfying the conditions of Theorem 5.1 is called a *reactance* or *Foster* matrix.

We now note that the conditions $k_{11} > 0$ and $k_{22} > 0$ are already known from Theorem 3.1 since z_{11} and z_{22} are driving-point impedances of lossless networks. This is clear from Fig. 5.1 if we remember that z_{11} is the input

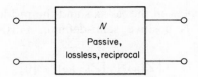

Fig. 5.1 A lossless reciprocal two-
port.

impedance of the lossless one-port obtained by open-circuiting the output
port. Similarly, z_{22} is the output impedance when the input is open-
circuited. Therefore z_{11} and z_{22} are reactance (Foster) functions. On the
other hand, condition (5.8b) is peculiar to two-ports and is termed the
residue condition. This condition implies that poles of z_{12} must also be poles
of z_{11} and z_{22}, because if $k_{11}k_{22}=0$ and $k_{12}\neq 0$ the residue condition (5.8b)
cannot be satisfied. However, it is possible for either z_{11} or z_{22} to have a
pole not shared by the other parameters. Such poles are called *private poles*
of z_{11} or z_{22}.

If the residue condition (5.8b) is satisfied with equality at *all poles*, the
z-parameters and the corresponding network are said to be *compact*.

5.3 SYNTHESIS OF A FOSTER IMPEDANCE MATRIX

We have, so far, established that for a given matrix $[Z]$ to be realizable as
the impedance matrix of a lossless reciprocal two-port, it is necessary that
$[Z]$ be a Foster matrix, i.e. satisfying the conditions of Theorem 5.1. It now
remains to demonstrate the sufficiency of these conditions by obtaining a
network capable of realizing any Foster matrix.

Let $[K]$ be one of the residue matrices $[K_0]$, $[K_i]$, or $[K_\infty]$. Also let $f(p)$ be
one of the three functions $1/p$, $2p/(p^2+\omega_i^2)$, or p. Accordingly, a typical term
in the expansion of the Foster matrix $[Z]$ in (5.6) is of the form

$$[Z_i]=[K]\cdot f(p) \tag{5.9}$$

We now *assume* that the above typical term is realizable as the network
shown in Fig. 5.2. In order that this assumption should be valid, we must now

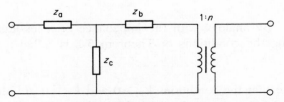

Fig. 5.2 A two-port capable of realizing any Foster
impedance matrix

demonstrate that it is *always* possible to find z_a, z_b, z_c as realizable reactance (Foster) functions, together with a real value of n, if $[K]$ is positive semi-definite.

Write the impedance matrix of the *assumed* network of Fig. 5.2 as

$$[Z_T] = \begin{bmatrix} z_a + z_c & nz_c \\ nz_c & n^2(z_b + z_c) \end{bmatrix} \tag{5.10}$$

Let us also assume that z_a, z_b, and z_c may be expresed as

$$z_a = af, \quad z_b = bf, \quad z_c = cf \tag{5.11}$$

where a, b, and c are positive constants. Equating $[Z_T]$ to a typical term in the expansion of $[Z]$ we obtain

$$\begin{bmatrix} k_{11}f & k_{12}f \\ k_{12}f & k_{22}f \end{bmatrix} = \begin{bmatrix} (a+c)f & ncf \\ ncf & n^2(b+c)f \end{bmatrix} \tag{5.12}$$

Thus, in order for the assumed network of Fig. 5.2 to be capable of realizing a typical term in the expansion of the given Foster matrix, we must have

$$(a+c) = k_{11}, \quad n^2(b+c) = k_{22}, \quad nc = k_{12} \tag{5.13}$$

From which, n has the same sign as k_{12}, and we have

$$c = \frac{k_{12}}{n} = \frac{|k_{12}|}{|n|} \tag{5.14a}$$

$$b = \frac{k_{22}}{n^2} - \frac{|k_{12}|}{|n|} \tag{5.14b}$$

$$a = k_{11} - \frac{|k_{12}|}{|n|} \tag{5.14c}$$

Therefore c is guaranteed non-negative. We must also require that a and b be non-negative, i.e.

$$k_{11} \geq \frac{|k_{12}|}{|n|}, \quad k_{22} \geq |n| \cdot |k_{12}| \tag{5.15}$$

or

$$\frac{|k_{12}|}{k_{11}} \leq |n| \leq \frac{k_{22}}{|k_{12}|} \tag{5.16}$$

which means that a value of n can always be found if

$$\frac{|k_{12}|}{k_{11}} \leq \frac{k_{22}}{|k_{12}|} \tag{5.17}$$

so that in the general case where the poles are common to z_{11}, z_{22}, and z_{12} we require

$$k_{11} > 0, \quad k_{22} > 0, \quad k_{11}k_{22} - k_{12}^2 \geq 0 \tag{5.18}$$

But the above conditions are precisely those of (5.8) which are satisfied by the most general residue matrices of the Foster matrix. Therefore, we conclude that it is always possible to find a network of the form shown in Fig. 5.2 with physically realizable components to realize a typical term in expansion (5.6) of the Foster matrix. Two-ports resulting from realizing the terms in (5.6) are then connected in series–series to obtain the complete realization of the given Foster matrix $[Z]$.

We have, therefore, established the *sufficiency* of the conditions of Theorem 5.1 for the realizability of a Foster matrix as a lossless reciprocal two-port. Moreover, examination of (5.14) to (5.16) reveals that a value of n

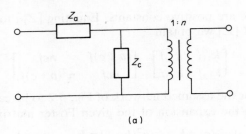

(a)

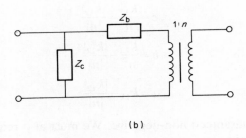

(b)

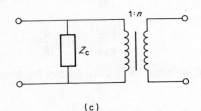

(c)

Fig. 5.3 Special cases of the two-port of Fig. 5.2. (a) $|n| = \dfrac{k_{22}}{|k_{12}|}$, (b) $|n| = \dfrac{|k_{12}|}{k_{11}}$, (c) The compact case: $|n| = \dfrac{|k_{12}|}{k_{11}} = \dfrac{k_{22}}{|k_{12}|}$.

may always be found to simplify the network of Fig. 5.2. Consider the following special cases:

(i)

$$|n| = \frac{k_{22}}{|k_{12}|} \tag{5.19}$$

so that from (5.14) we obtain $b = 0$, or $z_b = 0$ and the simplified network of Fig. 5.3(a) results. Alternatively we may have

$$|n| = \frac{|k_{12}|}{k_{11}} \tag{5.20}$$

so that from (5.14) we have $a = 0$, or $z_a = 0$ and the simplified network of Fig. 5.3(b) results.

(ii) If the residue condition (5.8b) is satisfied with equality, the network is *compact* and

$$|n| = \frac{|k_{12}|}{k_{11}} = \frac{k_{22}}{|k_{12}|} \tag{5.21}$$

Thus, from (5.14) we obtain $a = b = 0$, or $z_a = z_b = 0$. In this case the network reduces to the shunt z_c and the ideal transformer as shown in Fig. 5.3(c).

(iii) If the range of n in (5.16) includes unity, the choice $n = 1$ is possible and the transformer may be dispensed with.

(iv) For the compact case in (ii) and a typical network corresponding to $[K_\infty]$ or $[K_i]$ in (5.6), each network becomes that of Fig. 5.4(a). The cascade of the inductor and the ideal transformer may be replaced by a pair of perfectly coupled coils as explained in Chapter 2 (Fig. 2.2 and equations (2.7) to (2.11)). The transformation results in the network of Fig. 5.4(b).

Despite the above simplifications which may result, it must be noted that, in general, the synthesis of a lossless reciprocal two-port according to the technique presented here, requires the use of ideal transformers. The problem of transformerless canonic lossless two-port synthesis is, unfortunately, still unsolved.

We now give an alternative set of conditions for a given matrix to be a Foster matrix, and hence realizable as a lossless reciprocal two-port. These

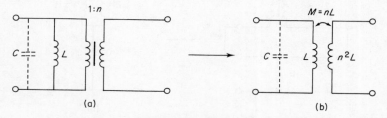

Fig. 5.4 A possible transformation in the compact case.

avoid the process of evaluating the residue matrices at the poles; therefore they can often be more convenient. These conditions follow directly from Theorem 5.1 and their proof is left as an exercise for the reader.

Corollary 5.1 *A 2×2 symmetric rational matrix $[Z(p)]$ is realizable as the impedance matrix of a lossless reciprocal two-port if and only if*

(a) z_{11}, z_{22}, and $\det [Z]/z_{22}$ are all reactance (Foster) functions
(b) z_{12} is an odd function of p.

Example 5.1 Verify that the following matrix is a Foster matrix and hence realize it as a lossless two-port

$$[Z(p)] = \begin{bmatrix} \dfrac{1+2p^2}{p} & \dfrac{3+p^2}{3p} \\ \dfrac{3+p^2}{3p} & \dfrac{18+p^2}{18p} \end{bmatrix}$$

Solution Expand $[Z]$ as

$$[Z] = \frac{1}{p}\begin{bmatrix} 1 & 1 \\ 1 & 1 \end{bmatrix} + p\begin{bmatrix} 2 & \frac{1}{3} \\ \frac{1}{3} & \frac{1}{18} \end{bmatrix}$$

$$= [Z_1] + [Z_2]$$

Hence $[Z]$ is expressible as in (5.6) with $[Z_1]$ and $[Z_2]$ satisfying conditions (5.8). In fact, the residue condition (5.8b) is satisfied with equality in both cases, hence the networks are compact.

For $[Z_1]$ we have $k_{11} = k_{22} = k_{12} = 1$. Therefore (5.16) gives $n = 1$ and (5.14) results in $a = b = 0$, or $z_a = z_b = 0$. Hence by reference to Fig. 5.3(c) we obtain the realization of $[Z_1]$ as a shunt 1 F capacitor, and the transformer can be dispensed with.

For $[Z_2]$ use of (5.21) gives $n = \frac{1}{6}$. From (5.14) we have $a = b = 0$, or $z_a = z_b = 0$, and $z_c = (k_{12}/n)p = 2p$. Hence z_c is a 2 H inductor.

The two networks realizing $[Z_1]$ and $[Z_2]$ are then connected in series–series to obtain the complete realization of $[Z]$ shown in Fig. 5.5(a). This

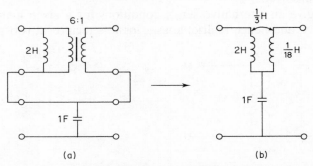

Fig. 5.5 Realization of the matrix of Example 5.1.

can be converted into the equivalent form of Fig. 5.5(b) which employs coupled coils, as explained in Chapter 2 by the aid of Fig. 5.4.

Example 5.2 Verify that the following matrix is a Foster matrix and find a realization.

$$[Z] = \begin{bmatrix} \dfrac{1+4p^2}{2p} & \dfrac{2-p^2}{4p} \\[2mm] \dfrac{2-p^2}{4p} & \dfrac{16+p^2}{32p} \end{bmatrix}$$

Solution   Expand $[Z]$ as

$$[Z] = \frac{1}{p}\begin{bmatrix} \frac{1}{2} & \frac{1}{2} \\ \frac{1}{2} & \frac{1}{2} \end{bmatrix} + p\begin{bmatrix} 2 & -\frac{1}{4} \\ -\frac{1}{4} & \frac{1}{32} \end{bmatrix}$$
$$= [Z_1] + [Z_2]$$

Hence $[Z]$ is expressible as in (5.6) with $[Z_1]$ and $\lfloor Z_2 \rfloor$ satisfying conditions (5.8). Therefore $[Z]$ satisfies the conditions of Theorem 5.1, and is a Foster matrix. Moreover the residue condition (5.8b) is satisfied with equality by both $[Z_1]$ and $[Z_2]$. Thus the network is compact.

For $[Z_1]$ we have $k_{11} = k_{22} = k_{12} = \frac{1}{2}$. Therefore either (5.16) or (5.21) gives $n = 1$, and (5.14) results in $a = b = 0$, $c = \frac{1}{2}$. Hence, by reference to Fig. 5.3(c) we obtain the realization of $[Z_1]$ as a shunt 2 F capacitor, and the transformer is dispensed with.

For $[Z_2]$ use of (5.16) or (5.21) gives $n = -\frac{1}{8}$. Also (5.14) gives $a = b = 0$, and $c = 2$. Hence z_c is a 2 H inductor.

The two networks realizing $[Z_1]$ and $[Z_2]$ are then connected in series-eries to obtain the complete realization of $[Z]$ shown in Fig. 5.6(a). We can also use the equivalence relations given in Chapter 2, equations (2.7) to (2.11) and Fig. 5.4 to convert the network of Fig. 5.6(a) to that of Fig. 5.6(b) which employs coupled coils.

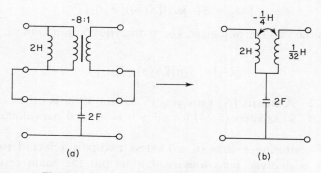

Fig. 5.6   Realization of the matrix of Example 5.2.

5.4 SYNTHESIS OF THE ADMITTANCE MATRIX OF A LOSSLESS RECIPROCAL TWO-PORT

The properties of the admittance matrix $[Y]$ of a lossless reciprocal two-port are identical to those derived for the impedance matrix. It is easy to show that $[Y]$ is a Foster matrix satisfying the conditions of Theorem 5.1, thus possessing an expansion of the same form as (5.6), i.e.

$$[Y] = \frac{[H_0]}{p} + 2p \sum_{i=1}^{n} \frac{[H_i]}{(p^2 + \omega_i^2)} + [H_\infty]p \qquad (5.22)$$

where $[H_0], [H_i], [H]$ are the residue matrices satisfying the same conditions in (5.8) with $[K]$ replaced by $[H]$. It is also possible to show that each term in (5.22) is realizable in the general form shown in Fig. 5.7. The individual two-ports are then connected in parallel–parallel to obtain the complete realization of (5.22). The details of this development are left to the reader.

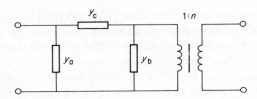

Fig. 5.7 A two-port capable of realizing any Foster admittance matrix.

5.5 THE SCATTERING MATRIX OF A LOSSLESS TWO-PORT

We have seen in Section 4.3 that the scattering matrix $[S(p)]$ of a passive two-port referred to strictly positive terminations is a bounded real matrix according to Definition 4.4. If in addition, the two-port is lossless, then Re $W = 0$ for $p = j\omega$ as a result of (2.34). Therefore (4.50) gives

$$[U_2] - [S(-j\omega)]'[S(j\omega)] = [0_2] \qquad (5.23)$$

where $[0_2]$ is the null (zero) matrix. Using analytic continuation we can write for all p.

$$[U_2] - [S(p)][S(p)] = [0_2] \qquad (5.24)$$

Definition 5.2 A matrix $[S]$ satisfying (5.23) for all ω is called unitary on the $j\omega$-axis. If $[S]$ satisfies (5.24) for all p it is termed para-unitary.

Thus, the scattering matrix of a lossless two-port referred to resistive terminations is *analytic para-unitary*. Let us put the main results for a lossless two-port in the form of a theorem.

Theorem 5.2 *The 2×2 scattering matrix $[S(p)]$ of any lossless two-port has the following properties*

(i) *The four entries in $[S(p)]$ are rational and real for p real.*
(ii) *$[S(p)]$ is analytic in Re $p \geq 0$, i.e. the four entries are analytic functions in the closed right half-plane (i.e. including the $j\omega$-axis, in particular the point at ∞).*
(iii) *$[S(j\omega)]$ is unitary, i.e. $[U_2] - [\tilde{S}(j\omega)][S(j\omega)] = [0_2]$. Thus $[S(p)]$ is para-unitary satisfying $[U_2] - [\underset{\sim}{S}(p)][S(p)] = [0_2]$ for all p.*
(iv) *If in addition to satisfying (i) to (iii), the matrix $[S]$ is symmetric $(S_{12} = S_{21})$, then the lossless two-port is reciprocal.*

Now, although the port normalizing numbers r_1 and r_2 in (4.20) to (4.24) are arbitrary positive resistors, in practice they are assigned the values of the actual terminations at the ports. As we have seen in Section 4.3, this results in the scattering parameters having meaningful physical interpretation in terms of the transmission of power from one port to the other and the reflection of power at a port. Consider the lossless two-port shown in Fig. 5.8, which has a scattering matrix $[S(p)]$ normalized to r_g at port 1 and r_ℓ at port 2. With port 2 terminated in r_ℓ and $V_g(p)$ driving port 1 we have form (4.45)

$$\frac{V_2(p)}{V_g(p)} = \frac{1}{2}\sqrt{\frac{r_\ell}{r_g}}S_{21}(p) \tag{5.25}$$

By definition, the transducer power gain $G(\omega^2)$ is the ratio of $P_\ell(\omega^2)$: the power delivered to the load r_ℓ, to $P_g(\omega^2)$: the power available from the generator. Thus from (5.25) and (4.46) we have,

$$G(\omega^2) = \frac{P_\ell(\omega^2)}{P_g(\omega^2)}$$
$$= |S_{21}(j\omega)|^2 \tag{5.26}$$

Clearly a lossless two-port inserted between a resistive source and a resistive load (Fig. 5.8) operates as a filter. The problem of designing a (reciprocal) lossless two-port to achieve a prescribed gain $|S_{21}(j\omega)|^2$ is one of *synthesis*. On the other hand, the problem of selecting a suitable realizable $S_{21}(j\omega)$ which satisfies certain optimality criteria is the subject of *approximation theory* which is discussed in a later chapter.

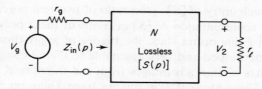

Fig. 5.8 The doubly terminated lossless two-port.

Now consider the para-unitary condition (5.24) rewritten as

$$[\underline{S}][S] = [U_2] \qquad (5.27)$$

where the argument p has been dropped for convenience. Write (5.27) explicitly as

$$\begin{bmatrix} S_{11*} & S_{21*} \\ S_{12*} & S_{22*} \end{bmatrix} \begin{bmatrix} S_{11} & S_{12} \\ S_{21} & S_{22} \end{bmatrix} = \begin{bmatrix} 1 & 0 \\ 0 & 1 \end{bmatrix} \qquad (5.28)$$

which gives

$$S_{11}S_{11*} + S_{21}S_{21*} = 1 \qquad (5.29)$$

$$S_{12}S_{11*} + S_{22}S_{21*} = 0 \qquad (5.30)$$

$$S_{11}S_{12*} + S_{21}S_{22*} = 0 \qquad (5.31)$$

$$S_{12}S_{12*} + S_{22}S_{22*} = 1 \qquad (5.32)$$

where (5.30) and (5.31) are the same with p replaced by $-p$. From (5.30) we have

$$S_{22} = -\frac{S_{11*}S_{12}}{S_{21*}} \qquad (5.33)$$

which when substituted in (5.32) gives

$$S_{12}S_{12*} + \frac{S_{11}S_{11*}S_{12}S_{12*}}{S_{21}S_{21*}} = 1 \qquad (5.34)$$

Using (5.29) in (5.34) we obtain

$$S_{12}S_{12*} = S_{21}S_{21*} \qquad (5.35)$$

and

$$S_{11}S_{11*} = S_{22}S_{22*} \qquad (5.36)$$

In particular, for $p = j\omega$, (5.35) and (5.36) become

$$|S_{12}(j\omega)| = |S_{21}(j\omega)| \qquad (5.37)$$

$$|S_{11}(j\omega)| = |S_{22}(j\omega)| \qquad (5.38)$$

In words, expression (5.37) means that the magnitude of the forward transmission coefficient S_{21} equals that of the reverse transmission coefficients S_{12} on the imaginary axis, *even for a non-reciprocal network*. Similarly from (5.38) the magnitudes of the input reflection coefficient S_{11} and the output reflection coefficient S_{22} are equal on the $j\omega$-axis.

Noting that each entry of $[S]$ is the ratio of two real polynomials, we now give a compact representation of $[S]$ in terms of these polynomials. Such a characterization is important since the theory of approximation by polynomials is highly developed and very amenable to digital computer implementation. It also has certain advantages from the theoretical viewpoint as we shall see later. This representation follows from Theorem 5.2 and is due to Belevitch.[20,28]

Corollary 5.2 (Belevitch Representation Theorem) *Any lumped lossless two-port N has the scattering matrix*

$$[S(p)] = \frac{1}{g(p)} \begin{bmatrix} h(p) & f(p) \\ \varepsilon f_*(p) & -\varepsilon h_*(p) \end{bmatrix} \qquad (5.39)$$

where

(a) $g(p), h(p),$ *and* $f(p)$ *are real polynomials;* $\varepsilon = \pm 1,$
(b) $g(p)$ *is strictly Hurwitz,*
(c)

$$h(p)h_*(p) + f(p)f_*(p) = g(p)g_*(p) \qquad (5.40)$$

If, in addition, N is reciprocal, then

$$[S(p)] = \frac{1}{g(p)} \begin{bmatrix} h(p) & f(p) \\ f(p) & -\varepsilon h_*(p) \end{bmatrix} \qquad (5.41)$$

where $f(p)$ *is either even or odd.* $\varepsilon = 1$ *if* $f(p)$ *is even and* $\varepsilon = -1$ *if* $f(p)$ *is odd,* ε *is called the* polarity *of* $f(p)$. *Thus, for a lossless reciprocal two-port*

$$f(p) = \varepsilon f_*(p), \qquad \varepsilon = \pm 1 \qquad (5.42)$$

and conditions (a) *to* (c) *are still satisfied. However,* (c) *may now be written as*

(d) $$h(p)h_*(p) + \varepsilon f^2(p) = g(p)g_*(p) \qquad (5.43)$$

It is to be noted that the representation (5.39) *allows common factors to exist between the numerator and denominator of any function.* For example f and g may possess a common factor. However, S_{12} is the *irreducible* form of f/g, i.e. the expression after the cancellation of the common factor. Similar remarks hold for the other entries.

We now derive the expressions for the impedance matrix in terms of the polynomials f, h, and g. The results are then used to prove the sufficiency of conditions (a), (b), and (d) of Corollary 5.2 for a reciprocal realization.

For a lossless two-port, not necessarily reciprocal, we substitute from (5.39) in (4.63) and make use of (5.40). This gives

$$\bar{z}_{11} = \frac{(g + \varepsilon g_*) + (h + \varepsilon h_*)}{(g - \varepsilon g_*) - (h - \varepsilon h_*)}$$

$$\bar{z}_{12} = \frac{2f}{(g - \varepsilon g_*) - (h - \varepsilon h_*)}$$

$$\bar{z}_{21} = \frac{2\varepsilon f_*}{(g - \varepsilon g_*) - (h - \varepsilon h_*)} \qquad (5.44)$$

$$\bar{z}_{22} = \frac{(g + \varepsilon g_*) - (h + \varepsilon h_*)}{(g - \varepsilon g_*) - (h - \varepsilon h_*)}$$

If the lossless two-port is reciprocal, then (5.42) gives $f = \varepsilon f_*$, i.e. f is either even or odd. Let

$$g_e = \text{Ev } g, \quad h_e = \text{Ev } h$$
$$g_0 = \text{Od } g, \quad h_0 = \text{Od } h \tag{5.45}$$

Thus, for a lossless reciprocal two-port expressions (5.44) reduce to

(i) $\varepsilon = 1$ (f even)

$$[\bar{Z}] = \frac{1}{(g_0 - h_0)} \begin{bmatrix} (g_e + h_e) & f \\ f & (g_e - h_e) \end{bmatrix} \tag{5.46}$$

(ii) $\varepsilon = -1$ (f odd)

$$[\bar{Z}] = \frac{1}{(g_e - h_e)} \begin{bmatrix} (g_0 + h_0) & f \\ f & (g_0 - h_0) \end{bmatrix} \tag{5.47}$$

Let us now assume that we are given a scattering matrix satisfying conditions (a), (b), and (d) of Corollary 5.2 for a *reciprocal lossless* two-port. Our objective is to show that these conditions *map directly* on to those of Corollary 5.1 for the impedance matrix, and are therefore also sufficient for realizability. First note that if $[\bar{Z}]$ as given in (5.46) or (5.47) is realizable as a lossless reciprocal two-port $\bar{N}$, then the denormalized impedance matrix $[Z]$ obtained from (4.60) to (4.61) is realizable as the same two-port but with a $\sqrt{r_g} : 1$ ideal transformer at the input and a $1 : \sqrt{r_\ell}$ ideal transformer at the output. This is illustrated in Fig. 5.9. Therefore we need only consider the normalized impedance matrix $[\bar{Z}]$.

We, therefore, are required to show that if the polynomials f, h, and g satisfy conditions (a), (b), and (d) of Corollary 5.2, then $[\bar{Z}]$ as obtained from (5.46) or (5.47) is a Foster matrix satisfying Corollary 5.1. This requires that $\bar{z}_{11}$, $\bar{z}_{22}$, and $\det[\bar{Z}]/\bar{z}_{22}$ are all Foster functions and $\bar{z}_{12}$ is an odd rational function.

First note that from (5.4), S_{11} is the *irreducible* form of h/g and from Theorem 5.2, S_{11} is a bounded real function. Therefore, using (2.65) the

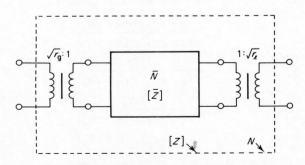

Fig. 5.9 Network interpretation of normalization and denormalization of the impedance matrix.

function

$$Z = \frac{1 + S_{11}}{1 - S_{11}}$$

$$= \frac{g + h}{g - h} \tag{5.48}$$

is a positive real function. Using (5.45) we can write Z as

$$Z = \frac{(g_e + h_e) + (g_0 + h_0)}{(g_e - h_e) + (g_0 - h_0)} \tag{5.49}$$

But from Corollary 2.9 the function obtained by interchanging the even (or odd) parts of the numerator and denominator of Z, is also p.r. Therefore

$$Z_1 = \frac{(g_e + h_e) + (g_0 - h_0)}{(g_e - h_e) + (g_0 + h_0)} \tag{5.50}$$

is a p.r.f. Using Corollaries 2.5 and 3.1, we may assert that the functions

$$\rho_1 = \frac{g_e + h_e}{g_0 + h_0}, \quad \rho_2 = \frac{g_e - h_e}{g_0 - h_0}$$

$$\rho_3 = \frac{g_e + h_e}{g_0 - h_0}, \quad \rho_4 = \frac{g_0 + h_0}{g_e - h_e} \tag{5.51}$$

are all Foster functions. Comparing the functions in (5.51) with the entries of $[\bar{Z}]$ in (5.46) and (5.47) it is evident that $\bar{z}_{11}$ and $\bar{z}_{22}$ as expressed in terms of h and g are guaranteed Foster functions. Furthermore, $\bar{z}_{12} = \bar{z}_{21}$ being the ratio of even-to-odd or odd-to-even polynomials. Thus $\bar{z}_{12}$ is an odd rational function. Also from (5.46) we obtain

$$\det [\bar{Z}] = \frac{(g_e^2 - h_e^2) - f^2}{(g_0 - h_0)^2}$$

$$= \frac{(g + g_*)^2 - (h + h_*)^2 - 4f^2}{4(g_0 - h_0)^2}$$

and use of (5.43) with $\varepsilon = 1$ gives

$$\det [\bar{Z}] = \frac{g_0 + h_0}{(g_0 - h_0)}$$

Hence

$$\frac{\det [\bar{Z}]}{\bar{z}_{22}} = \frac{g_0 + h_0}{g_e - h_e} = \rho_4$$

and is, therefore, a Foster function. Similarly we can show that (5.47) leads to $\det [\bar{Z}]/\bar{z}_{22} = \rho_3$, a Foster function.

From the above results, it follows that the properties of $[S]$ as stated in (a), (b), and (d) of Corollary 5.2 (or indeed Theorem 5.2) for a *lossless reciprocal* two-port *map directly* on to those of $[\bar{Z}]$ as dictated by Corollary 5.1 (or Theorem 5.1). Therefore, we have the following result.

Theorem 5.3 A *symmetric matrix*

$$[S(p)] = \frac{1}{g(p)} \begin{bmatrix} h(p) & f(p) \\ f(p) & -\varepsilon h_*(p) \end{bmatrix}$$

is the scattering matrix of a lumped lossless reciprocal two-port if and only if

(i) $f(p)$, $h(p)$, *and* $g(p)$ *are real polynomials.*
(ii) $f(p) = \varepsilon f_*(p)$, *i.e.* $f(p)$ *is either even or odd.* $\varepsilon = 1$ *for* $f(p)$ *even and* $\varepsilon = -1$ *for* $f(p)$ *odd.*
(iii) $g(p)$ *is strictly Hurwitz.*
(iv) $h(p)h_*(p) + \varepsilon f^2(p) = g(p)g_*(p)$.

It should be noted that Theorem 5.3 establishes the *necessity and sufficiency* of the given conditions *specially* for a *reciprocal* realization, since we relied on the results of Theorem 5.1 in Section 5.3. However, it can be shown[20] that conditions (i) to (iii) of Theorem 5.2 or conditions (a) to (c) of Corollary 5.2, are both necessary and sufficient for a *non-reciprocal* realization containing *gyrators*; although of course we have only established the necessity of these conditions.

5.6 THE TRANSMISSION MATRIX OF A LOSSLESS TWO-PORT

We now give the expressions for the transmission matrix of a lossless two-port in terms of the polynomials f, h, and g in the scattering description. Again it is only necessary to consider the normalized transmission matrix $[\bar{T}]$ related to $[T]$ by (4.76). This is because if $[\bar{T}]$ is realizable as lossless two-port $\bar{N}$, then $[T]$ is realizable as the same network but with a $\sqrt{r_g}:1$ ideal transformer at the input and a $1:\sqrt{r_\ell}$ ideal transformer at the output. Figure 5.9 still shows the two networks. To obtain $[\bar{T}]$ in terms of h, f, and g we can either substitute from (5.39) into (4.78), or use expressions (5.44) for $[Z]$ together with (4.72).

Thus, for a lossless two-port, *not necessarily reciprocal*, we have

$$[\bar{T}(p)] = \frac{1}{\varepsilon f_*}[t(p)] \tag{5.52}$$

where $[t(p)]$ is a *polynomial matrix* expressed as

$$[t(p)] = \begin{bmatrix} a(p) & b(p) \\ c(p) & d(p) \end{bmatrix}$$

$$= \frac{1}{2}\begin{bmatrix} (g+\varepsilon g_*)+(h+\varepsilon h_*) & (g-\varepsilon g_*)+(h-\varepsilon h_*) \\ (g-\varepsilon g_*)-(h-\varepsilon h_*) & (g+\varepsilon g_*)-(h+\varepsilon h_*) \end{bmatrix} \tag{5.53}$$

Using the properties of f, h, and g as given in Corollary 5.2, we can easily obtain the corresponding properties of $[t(p)]$. From (5.53) we note that if $\varepsilon = 1$, then $a(p)$ and $d(p)$ are even polynomials while $b(p)$ and $c(p)$ are odd. Conversely for $\varepsilon = -1$, $a(p)$ and $d(p)$ are odd whereas $b(p)$ and $c(p)$ are even. Moreover, adding all the entries of $t(p)$ we obtain $g(p)$ which is strictly Hurwitz by condition (b) of Corollary 5.2. Also the function

$$F(p) = \frac{a(p) + b(p)}{c(p) + d(p)}$$

$$= \frac{g(p) + h(p)}{g(p) - h(p)} \tag{5.54}$$

is a p.r.f. by comparison with (5.48). Next evaluate

$$\det[t(p)] = ad - bc$$

$$= \varepsilon(gg_* - hh_*) \tag{5.55}$$

and use of (5.40) gives

$$\det[t(p)] = \varepsilon ff_* \tag{5.56}$$

In the *reciprocal* case, the conditions obtained so far are still in force, but (5.42) gives $f = \varepsilon f_*$ and (5.56) becomes

$$\det[t(p)] = f^2(p) \tag{5.57}$$

which means that $\det[t(p)]$ is the *square of an even or odd polynomial for a lossless reciprocal two-port*. Hence we have the following results.

Theorem 5.4 *The most general passive lumped, lossless two-port has the 2×2 real polynomial transmission matrix*

$$[t(p)] = \begin{bmatrix} a(p) & b(p) \\ c(p) & d(p) \end{bmatrix}$$

with the following properties

(i) *$a(p)$ and $d(p)$ are both even or both odd.*
(ii) *$b(p)$ and $c(p)$ are both odd if $a(p)$ is even, and both even if $a(p)$ is odd.*
(iii) *$a(p) + b(p)$ and $c(p) + d(p)$ are relatively prime in Re $p \geq 0$.*
(iv) *$a(p) + b(p) + c(p) + d(p)$ is strictly Hurwitz.*
(v) *$a(p)d(p) - b(p)c(p) = \varepsilon f(p)f_*(p)$.*

Corollary 5.3 *$[t(p)]$ is the polynomial transmission matrix of a lumped lossless reciprocal two-port, if and only if it satisfies the conditions of Theorem 5.4 with $f_* = \varepsilon f$, i.e. f is either even or odd and instead of (v) we have*

(v)′ $$a(p)d(p) - b(p)c(p) = f^2$$

The sufficiency of the conditions of Corollary 5.3 (for the reciprocal case) can be easily established by showing that these lead to the decomposition in Theorem 5.3 with the required properties. Although the non-reciprocal case is of little interest to us, it may be shown that the conditions (i) to (v) of Theorem 5.4 are also sufficient for a non-reciprocal realization containing gyrators.

5.7 THE DOUBLY TERMINATED LOSSLESS TWO-PORT

5.7.1 The central problem in filter synthesis

Networks are designed to perform specific tasks! The theoretical development in this book leads naturally to the design techniques of filters and related networks. The central problem in passive filter design consists in the synthesis of a lossless two-port to operate between a resistive source and a resistive load as shown in Fig. 5.8; the realized network meets a prescribed set of specifications regarding its amplitude and/or phase characteristics. These specifications are expressed mathematically in the form of a transfer function $S_{21}(p)$ with the required properties. The derivation of a realizable $S_{21}(p)$ such that the corresponding realized network gives the required response, is the subject of *approximation theory*, to be discussed in a later chapter.

Although, initially, our study concentrates on the specific case of a passive network, the fundamental network-theoretic ideas which are developed for this case form the essential background for the other categories to be discussed later, including active ones.

5.7.2 The zeros of transmission

Consider any lossless two-port, not necessarily reciprocal. With port 1 excited and port 2 terminated in r_ℓ as shown in Fig. 5.8, any zero p_0 of $S_{21}(p)$ in Re $p \geq 0$ is a zero of transmission (or transmission zero), i.e. under the excitation $e^{p_0 t}$ the current in r_ℓ (hence the power) is zero. This follows directly from (5.25). Similarly, with port 2 excited and port 1 terminated in r_g, the zeros of $S_{12}(p)$ in Re $p \geq 0$ are the zeros of transmission in the opposite direction. But from the decomposition in (5.39) S_{21} is the irreducible form of $\varepsilon f_*/g$ and S_{12} is the irreducible form of f/g. Hence the zeros of transmission in Re $p \geq 0$ are those of the irreducible form of ff_*/g^2. We also have from (5.40)

$$\text{Ir} \frac{ff_*}{g^2} = \text{Ir} \frac{(gg_* - hh_*)}{g^2} \tag{5.58}$$

where Ir stands for 'irreducible form of', i.e. the expression after the cancellation of possible common factors between numerator and denominator. Note that cancellation of zeros in the above function may only occur at the zeros of g which are zeros of h or h_*. Now ff_* is a real even polynomial, therefore its zeros are distributed symmetrically with respect to the $j\omega$-axis and are double on this axis. On the other hand, g is strictly

Hurwitz by Corollary 5.2, so that the cancelled zeros in (5.58) must be in the *open left half-plane*: Re $p < 0$. Although the zeros of (5.28) on the $j\omega$-axis occur automatically with even multiplicity, it is the *convention to count them for half their order*. Consequently the number of finite zeros of transmission in Re $p \geq 0$ is the sum of the zeros of f in Re $p \geq 0$ and those of f_* in Re $p > 0$. Thus the number of finite zeros of transmission m_z in Re $p \geq 0$ is equal to the degree of f, i.e.

$$m_z = \deg f(p) \tag{5.59}$$

The number of zeros of transmission at infinity is

$$m_\infty = \deg g(p) - \deg f(p) \tag{5.60}$$

Hence the total number of zeros of transmission in Re $p \geq 0$ including those at infinity is given by

$$m = m_z + m_\infty \tag{5.61}$$

i.e.

$$m = \deg g(p) \tag{5.62}$$

In the reciprocal case (5.42) gives $f = \varepsilon f_*$, i.e. f is either even or odd, and we have

$$ff_* = \varepsilon f^2 \tag{5.63}$$

therefore each distinct zero of transmission occurs with even multiplicity. Thus, a finite zero of transmission of a *lossless reciprocal* two-port is a double-order zero.

Now consider the input impedance $Z_{in}(p)$ of the resistor-terminated lossless two-port, not necessarily reciprocal, of Fig. 5.8. Let the input reflection coefficient of the network be $S_{11}(p)$ referred to $r_g = 1\ \Omega$ for convenience. Z_{in} and S_{11} are related by (2.74), i.e.

$$Z_{in} = \frac{1 + S_{11}}{1 - S_{11}}$$

$$= \frac{N}{D} \tag{5.64}$$

where, using the polynomial representation of S_{11} in (5.39) we have

$$Z_{in} = \text{Ir}\ \frac{g + h}{g - h} \tag{5.65}$$

Evaluating the even-part of Z_{in} according to

$$\text{Ev}\ Z_{in} = \tfrac{1}{2}\ \text{Ir}\ (Z_{in} + Z_{in*})$$

$$= \tfrac{1}{2}\ \text{Ir}\ \left\{ \frac{ND_* + N_*D}{DD_*} \right\} \tag{5.66}$$

and using (5.64) together with (5.40) we obtain

$$\text{Ev } Z_{\text{in}} = \tfrac{1}{2} \text{Ir} \left\{ \frac{gg_* - hh_*}{(g-h)(g_* - h_*)} \right\}$$

$$= \tfrac{1}{2} \text{Ir} \left\{ \frac{ff_*}{(g-h)(g_* - h_*)} \right\} \tag{5.67}$$

Comparing (5.67) with (5.58) we conclude that Ev Z_{in} contains all the zeros of transmission unless there is some additional cancellation in (5.67) which does not occur in (5.58). This can be the case if a zero of transmission p_0 is *also a pole* of Z_{in} in (5.65). But at such a point, we have

$$Z_{\text{in}*}(p_0) = Z_{\text{in}}(-p_0) = \text{Ir} \left. \frac{g_* + h_*}{g_* - h_*} \right|_{p=p_0}$$

$$= \text{Ir} \left. \frac{g_* h + h h_*}{g_* h - h h_*} \right|_{p=p_0}$$

and use of (5.40) gives

$$Z_{\text{in}}(-p_0) = \text{Ir} \left. \frac{g_*(g+h) + ff_*}{g_*(h-g) - ff_*} \right|_{p=p_0}$$

$$= \text{Ir} \left. \frac{g+h}{h-g} \right|_{p=p_0}$$

Hence, if p_0 is a zero of transmission and simultaneously a pole of Z_{in}, then $-p_0$ is also a pole of Z_{in}. But this is *impossible* for a *complex* p_0 since either p_0 or $-p_0$ would have to lie in the open right half-plane, which contradicts the positive real character of Z_{in}. Therefore, this pole must be on the $j\omega$-axis.

The above discussion also shows that if the lossless two-port is described by its polynomial transmission matrix $[t(p)]$ defined in (5.53), then the finite zeros of transmission are also contained in det $[t(p)]$. This follows directly by comparison of (5.58) and (5.56).

We now put the above results in the form of a theorem for later reference.

Theorem 5.5 *Let N be a lossless two-port, then the following is the case:*

(i) *If N is described by its scattering matrix as expressed in (5.39) then the zeros of transmission are the zeros of the irreducible form of ff_*/g^2. If N is reciprocal then $f = \varepsilon f_*$, i.e. f is either even or odd, and the zeros of transmission are those of the irreducible form of f^2/g^2, hence they are of even multiplicity.*

(ii) *Let the input impedance of the resistor-terminated lossless two-port be $Z_{\text{in}} = N/D$. Then Ev Z_{in} as expressed in (5.66) contains all the zeros of transmission except those which are also $j\omega$-axis poles of Z_{in}. On the other* hand, the polynomial

$$u(p) = ND_* + N_*D$$

contains all the finite zeros of transmission, including those which are simultaneously $j\omega$-axis poles of Z_{in}.

(iii) If N is described by its polynomial transmission matrix $[t(p)]$ as expressed in (5.53) then the finite zeros of transmission are those of

$$\det [t(p)] = ad - bc.$$

5.7.3 Symmetry and antimetry

If a one-port is formed from a lossless two-port by closing its output on a resistor as shown in Fig. 5.10(a) then the input impedance is related to the input reflection coefficient by (2.74), i.e.

$$Z_{\text{in}} = r_g \frac{1 + S_{11}}{1 - S_{11}} \qquad (5.68)$$

Similarly for the output impedance when the input is closed on r_g as shown in Fig. 5.10(b),

$$Z_{\text{out}} = r_\ell \frac{1 + S_{22}}{1 - S_{22}} \qquad (5.69)$$

The network N is said to be *electrically symmetric with respect to the termination* if

$$\frac{Z_{\text{in}}}{Z_{\text{out}}} = \frac{r_g}{r_\ell} \qquad (5.70)$$

while N is said to be *electrically antimetric* if

$$Z_{\text{in}}Z_{\text{out}} = r_g r_\ell \qquad (5.71)$$

Using (5.68) and (5.69) in (5.70), symmetry requires

$$S_{11} = S_{22} \qquad (5.72a)$$

or in terms of the polynomial decomposition in (5.39)

$$h = -\varepsilon h_* \qquad (5.72b)$$

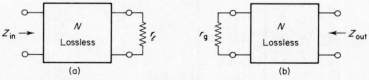

Fig. 5.10   Pertinent to the definitions of symmetry and antimetry.

Similarly N is antimetric if

$$S_{11} = -S_{22} \tag{5.73a}$$

or

$$h = \varepsilon h_* \tag{5.73b}$$

It follows that for a symmetric or antimetric network, $h = \pm h_*$, i.e. h is either even or odd.

PROBLEMS

5.1 Verify that the following is a Foster impedance matrix and find a realization

$$[Z(p)] = \begin{bmatrix} \dfrac{1+p^2}{2p} & \dfrac{4+p^2}{8p} \\ \dfrac{4+p^2}{8p} & \dfrac{16+p^2}{32p} \end{bmatrix}$$

5.2 Realize the Foster matrix,

$$[Z(p)] = \begin{bmatrix} \dfrac{p^4+3p^2+1}{p^3+p} & \dfrac{2p^4+2p^2+1}{p^3+p} \\ \dfrac{2p^4+2p^2+1}{p^3+p} & \dfrac{4p^4+6p^2+1}{p^3+p} \end{bmatrix}$$

and show that the resulting lossless two-port is compact.

5.3 Evaluate the scattering matrix, referred to $1\,\Omega$ terminations, for each of the lossless two-ports of Problems 5.1 and 5.2.

5.4 Consider a passive lossless n-port as shown in Fig. P5.4. Show that it possesses a scattering matrix $[S]$, referred to real terminations, which satisfies the same properties of Theorem 5.2 except that $[S]$ is now an $n \times n$ matrix. In particular, show that the entries of $[S]$ are defined by

(a) S_{ii} = reflection coefficient of the one-port formed by closing all the other ports on their reference resistors.

(b) $S_{ij}(i \neq j)$ = transmission coefficient from port j to port i under reference terminating conditions.

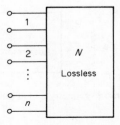

Fig. P5.4   Problem 5.4.

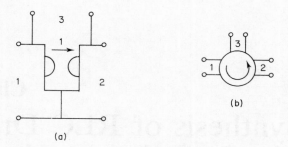

Fig. P5.5 Problem 5.5. (a) Forming a circulator from
a gyrator, (b) symbol of a three-port ideal circulator.

5.5 Consider the non-reciprocal three-port shown in Fig. P5.5(a). This is formed by adding a pair of terminals to an ideal gyrator. Find its scattering matrix referred to $1\,\Omega$ resistors and show that the main diagonal entries are zero. Also show that when a signal is incident at port i $(i = 1 \rightarrow 3)$ with all ports matched, none is reflected, none is transmitted to port $(i-1)$ but all the signal is transmitted to port $(i+1)$ without loss. Thus, the three-port has a *cyclic power transmission property* and is called a three-port *circulator*. The symbol for a circulator is hown in Fig. P5.5(b).

Chapter 6

Synthesis of RLC Driving-point Impedances: the Link with Filter Synthesis

6.1 INTRODUCTION

We are now in a position to discuss the synthesis of an arbitrary positive real impedance $Z(p)$. This is the most general type of realizable function. We have seen in Chapter 2 (Theorem 2.1) that for $Z(p)$ to be the driving-point impedance of a passive one-port, it is *necessary* that $Z(p)$ be a positive real function (p.r.f.). We now ask the question: is the p.r. condition also *sufficient*? In other words, given an arbitrary impedance that satisfies nothing more (or less) than the p.r. condition, is it always possible to realize the impedance as the driving-point impedance of a passive one-port? It will be shown that the answer is in the affirmative.

We first discuss the *global* synthesis, proving the sufficiency of the p.r. condition on $Z(p)$ for realizability as a passive one-port containing the conventional circuit elements. Next, the details of the synthesis technique are given in the form of a general theory of cascade synthesis. The development in this chapter establishes, firmly and unequivocally, the link between the theoretical exposition in this book, and the filter design problem.

Historically, the first proof of the sufficiency of the p.r. condition for realizability was given by Brune.[29] However, it was Darlington's demonstration that made network synthesis the powerful and rigorous discipline it is today. In his classic paper,[30] Darlington showed that any p.r.f. can be realized as the driving-point impedance of a *lossless reciprocal two-port, terminated in a* (1Ω) *resistor* as shown in Fig. 6.1. This result has come to be known as *Darlington's Theory*. It constitutes an existence theorem demonstrating the sufficiency of the p.r. condition for realizability. However, the main significance of Darlington's result lies in that it reduces the synthesis of a p.r.f. to that of a lossless reciprocal two-port; this being the central problem in filter synthesis.

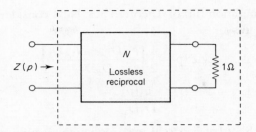

Fig. 6.1 Realization of an arbitrary positive
real impedance.

6.2 SUFFICIENCY OF THE POSITIVE REAL CONDITION

Given a rational p.r.f. $Z(p)$ it is required to show that it is always realizable
as the driving-point impedance of a passive one-port containing the conventional elements. In particular, we demonstrate that, relying only on the p.r.
condition, $Z(p)$ is realizable as the driving-point impedance of a lossless
reciprocal two-port terminated in a $(1 \, \Omega)$ resistor, as shown in Fig. 6.1. The
conditions for $Z(p)$ to be positive real were given in Chapter 2: Theorem
2.1 or Corollary 2.8.

Let the given p.r.f. $Z(p)$ be written as

$$Z(p) = \text{Ir} \, \frac{N}{D} = \frac{N_1}{D_1} \tag{6.1}$$

where, as before, 'Ir' stands for 'irreducible form of'. Thus N_1 and D_1 are
relatively prime. The input reflection coefficient, relative to $1 \, \Omega$, say, of the
one-port whose driving-point impedance is $Z(p)$ is given by

$$S_{11}(p) = \frac{Z(p) - 1}{Z(p) + 1}$$

$$= \text{Ir} \, \frac{h}{g} = \frac{h_1}{g_1} \tag{6.2}$$

where $h = N - D$, $g = N + D$, and h_1 and g_1 are relatively prime.

It is now required to show that $Z(p)$ can be identified as the input
impedance of the $1 \, \Omega$-terminated lossless reciprocal two-port N shown in
Fig. 6.1. To this end, we must show that the p.r. condition on $Z(p)$ leads to a
scattering description of the two-port N which satisfies the conditions of
Theorem 5.3, or conditions (a), (b), and (d) of Corollary 5.2.

First note that since $Z(p)$ is a p.r.f., then both N and D are real
polynomials; consequently h_1 and g_1 are also real polynomials, hence they
satisfy condition (i) of Theorem 5.3. Secondly, by Corollary 2.8 the sum
$(N_1 + D_1)$ is strictly Hurwitz. Hence g_1 in (6.2) is strictly Hurwitz as

demanded by condition (iii) of Theorem 5.3. Next consider the even-part of $Z(p)$

$$\text{Ev } Z(p) = \frac{1}{2} \left\{ \frac{N_1 D_{1*} + N_{1*} D_1}{D_1 D_{1*}} \right\}$$

$$= \frac{u_1(p)}{D_1 D_{1*}} \tag{6.3}$$

By Corollary 2.8, $u_1(p)$ is a real even polynomial, non-negative along the entire $j\omega$-axis, expressible as

$$u_1(p) = f_1 f_{1*} \tag{6.4}$$

Substituting from (6.2) into (6.3) we have

$$u_1(p) = g_1 g_{1*} - h_1 h_{1*} \tag{6.5}$$

Therefore

$$g_1 g_{1*} - h_1 h_{1*} = f_1 f_{1*} \tag{6.6}$$

and f_1 is a real polynomial. Therefore the decomposition in (5.39) is possible. But f_1 is not unique, since it is obtained by factoring $(g_1 g_{1*} - h_1 h_{1*})$ and arbitrarily assigning half the zeros to f_1 and the other half to f_{1*}. However, by the p.r. condition (see the discussion of the factors of $u_1(p)$ in 2.11) the zeros of $u_1(p)$ on the $j\omega$-axis are of even multiplicity. Therefore $f_1 f_{1*} = u_1$ may be written as

$$u_1(p) = f_1 f_{1*} = \theta \theta_* \gamma^2 \tag{6.7}$$

where θ contains all the zeros in Re $p < 0$, θ_* contains the images of these zeros in Re $p > 0$, and γ^2 contains all the (double) $j\omega$-axis zeros.

Putting together the above results, we conclude that the polynomials h_1, f_1, and g_1 satisfy the conditions of Corollary 5.2 for a lossless two-port, *not necessarily reciprocal*. But there is *nothing* in the p.r. condition on an *irreducible* impedance to guarantee that $f_1 = \varepsilon f_{1*}$ ($\varepsilon = \pm 1$) and $u_1 = \varepsilon f_1^2$, f_1 being even or odd as demanded by condition (ii) of Theorem 5.3 for a *reciprocal* lossless two-port. However, this difficulty can be resolved in a straightforward manner. Let $Z(p)$ be *augmented* as follows

$$Z(p) = \frac{N_1 Q}{D_1 Q} = \frac{N}{D} \tag{6.8}$$

where Q is a strictly Hurwitz polynomial, to be chosen such that all the zeros of Ev $Z(p)$ are of even multiplicity. *Note that the new $Z(p)$ is no longer irreducible.* Clearly the input reflection coefficient becomes

$$S_{11}(p) = \frac{h_1}{g_1} \cdot \frac{Q}{Q} = \frac{h}{g} \tag{6.9}$$

Again we calculate

$$\text{Ev } Z(p) = \frac{u(p)}{DD_*} \tag{6.10}$$

where

$$2u(p) = ND_* + N_*D \tag{6.11}$$

Using (6.4) to (6.10) the new polynomial is given by

$$u(p) = gg_* - hh_*$$
$$= (g_1 g_{1*} - h_1 h_{1*})QQ_* \tag{6.12}$$

Therefore, by (6.6) and (6.7) we obtain

$$u(p) = \theta\theta_* QQ_* \gamma^2$$
$$= ff_* \tag{6.13}$$

so that in order to make $f = \varepsilon f_*$ and $u = \varepsilon f^2$, f being even or odd, we must choose

$$Q = \theta, \quad Q_* = \varepsilon\theta_* \tag{6.14}$$

which results in

$$u(p) = (gg_* - hh_*) = \varepsilon f^2 \tag{6.15}$$

where

$$f = QQ_* \gamma \tag{6.16}$$

as demanded by condition (ii) of Theoem 5.3. Clearly condition (iv) is also satisfied. The effect of the augmentation process (i.e. introducing the common factor Q in Z) on the scattering matrix of the lossless two-port can be summarized by

$$\frac{1}{g_1}\begin{bmatrix} h_1 & f_1 \\ \varepsilon f_{1*} & -\varepsilon h_{1*} \end{bmatrix} \to \frac{1}{g_1 Q}\begin{bmatrix} h_1 Q & QQ_*\gamma \\ QQ_*\gamma & -\varepsilon h_{1*}Q_* \end{bmatrix}$$
$$\to \frac{1}{g}\begin{bmatrix} h & f \\ f & -\varepsilon h_* \end{bmatrix} \tag{6.17}$$

Hence, the new set of augmented polynomials h, f, and g satisfy the conditions of Theorem 5.3 for a *lossless reciprocal* two-port.

It is worth noting that the effect of augmenting Z by Q/Q on the transmission coefficients S_{12} and S_{21} of the lossless two-port N is as follows

$$S_{12} = \frac{f_1}{g_1} \to S_{12} \cdot \frac{Q}{Q} \tag{6.18}$$

$$S_{21} = \frac{\varepsilon f_{1*}}{g_1} \to S_{21} \cdot \varepsilon \frac{Q_*}{Q} \tag{6.19}$$

Equivalently, S_{12} is augmented by Q_*/Q and S_{21} by Q/Q. The new product $S_{12}S_{21}$ has the excess factor QQ_* in its numerator, which allows the identification in (6.16). A function of the form

$$\pm \frac{Q_*}{Q} = \pm \frac{Q(-p)}{Q(p)} \tag{6.20}$$

is termed an *all-pass* function, since its magnitude is unity for all $p = j\omega$,

$$\left| \frac{Q(-j\omega)}{Q(j\omega)} \right| = 1 \tag{6.21}$$

To summarize, *the augmentation of $Z(p)$ consists in introducing a common factor $Q(p)$ between numerator and denominator, such that $Q(p)$ contains all the zeros in* Re $p < 0$ *of odd multiplicity in the even part of the impedance.* Therefore all the zeros of transmission become of even multiplicity as required for a lossless reciprocal two-port. Since augmenting the impedance is always possible, we have in fact demonstrated the *sufficiency of the p.r. condition* for the realizability of $Z(p)$ as shown in Fig. 6.1.

Theorem 6.1 (Darlington's Theory) *A given rational function $Z(p)$ is realizable as the driving-point impedance of a passive lumped lossless reciprocal two-port terminated in a $(1\ \Omega)$ resistor as shown in Fig. 6.1, if and only if $Z(p)$ is a positive real function.*

Having obtained the polynomials h, f, and g from the given impedance, the actual synthesis may, in principle, be performed by evaluating the impedance matrix of the lossless two-port from (5.46) or (5.47), then use is made of the synthesis technique of Chapter 5 (Section 5.3). Alternatively, the admittance matrix may be evaluated and the technique outlined in Section 5.4 is employed. Although this method is quite general, it suffers from two main practical disadvantages. First, it requires a large number of ideal transformers in a high degree network. Secondly, the effect of the variation of any particular element value on the overall response of the network cannot be easily isolated. Later in the present chapter an alternative technique of cascade synthesis will be given, which attempts to eliminate the drawbacks of the earlier technique of Chapter 5.

6.3 RELATION TO THE FILTER DESIGN PROBLEM

6.3.1 A return to the doubly terminated lossless two-port

So far, it has been assumed that the function $Z(p)$ to be realized is *given*. However, this is hardly the case in practice. Instead, *we are also required to find $Z(p)$ such that the realized network performs a certain task*. Throughout this book, we are mainly interested in the design of lossless two-ports to operate between a resistive source and a resistive load, as well as other categories which are modelled after these doubly terminated lossless two-

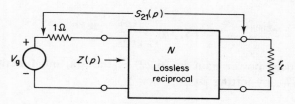

Fig. 6.2 The doubly-terminated lossless reciprocal two-port.

ports. The specifications on the performance of the network are normally interpreted as properties of the transmission coefficient $S_{21}(p)$ or its magnitude-squared $|S_{21}(j\omega)|^2$. Either of these is derived mathematically, and the derivation in the cases of filters and closely related networks is treated in later chapters on *approximation theory*.

Consider the doubly terminated lossless two-port shown in Fig. 6.2. Let us assume that the synthesis process takes, as a starting point, the transfer function $S_{21}(p)$ or the transducer power gain $|S_{21}(j\omega)|^2$. Either of these functions defines the required properties of the network to be realized. It is now required to evaluate the input impedance $Z(p)$ of the resistor-terminated lossless two-port. First note that we can always assume that the source resistance $r_g = 1\,\Omega$, since the realized network can be later impedance-scaled by the actual value r_g at the final stage of the synthesis. Secondly, for many types of filter transfer functions, the lossless two-port must operate between equal source and load resistors. In such cases, the choice $r_g = r_\ell = 1\,\Omega$ can be made and, again, we can scale the impedances in the realized network. However, certain types of response cannot be achieved with equal terminations unless ideal transformers are used. Nevertheless, the values $r_g = r_\ell = 1$ can always be assumed, since by a combination of impedance scaling and ideal transformers, any values of r_g and r_ℓ can be accommodated. In the following analysis we assume, as in Fig. 6.2, that $r_g = 1\,\Omega$, but the load is an arbitrary resistor r_ℓ.

6.3.2 Derivation of the input impedance from partial specification of the scattering matrix

We now give the procedure for evaluating $Z(p)$. The starting function can be either of the following

(a)
$$|S_{21}(j\omega)|^2 = \frac{|f(j\omega)|^2}{|g(j\omega)|^2} \tag{6.22}$$

or

(b)
$$S_{21}(p) = \frac{\varepsilon f_*(p)}{g(p)} \tag{6.23a}$$

thus

$$S_{12}(p) = \frac{f(p)}{g(p)} \tag{6.23b}$$

If the transducer power gain in (6.22) is the starting function, then we evaluate $S_{21}S_{21*}$ by letting $j\omega \to p$, i.e.

Step (1)

$$\frac{ff_*}{gg_*} = \left\{ \frac{|f(j\omega)|^2}{|g(j\omega)|^2} \right\}_{\omega \to p/j} \tag{6.24}$$

Then $gg_*(\equiv g(p)g(-p))$ is factored and the *open left half-plane zeros* are assigned to $g(p)$ since it must be *strictly* Hurwitz.

If, on the other hand, $S_{21}(p)$ or $S_{12}(p)$ in (6.23) is the starting function, then we proceed directly to the next step.

Step (2) Using (5.40) we evaluate

$$hh_* = gg_* - ff_* \tag{6.25}$$

Step (3) $hh_*(\equiv h(p)h(-p))$ is factored to obtain $h(p)$. This factorization is, of course, not unique.

Step (4) From $h(p)$ as obtained in Step (3), and $g(p)$ as given directly in (6.23) or obtained in Step (1), we evaluate the input impedance of the resistor-terminated lossless two-port as

$$Z(p) = \frac{g(p) + h(p)}{g(p) - h(p)} \tag{6.26}$$

Step (5) If in (6.24) the function ff_* is not the square of an even or odd polynomial, then $Z(p)$ must be augmented as explained in (6.8) for a reciprocal realization,

$$Z(p) = \frac{g(p) + h(p)}{g(p) - h(p)} \cdot \frac{Q(p)}{Q(p)} \tag{6.27}$$

with $Q(p)$ containing all the zeros in $\operatorname{Re} p < 0$ of odd multiplicity of ff_*.

A slight modification of the above procedure is possible, in the case where f is a general polynomial, i.e. neither even nor odd. Instead of going through the Steps (1) to (4) then augmenting Z in Step (5), *it is possible to augment $S_{12} = f/g$ itself at the outset.* This can be done by reference to expressions (6.17) to (6.20), from which it is evident that augmenting Z by Q/Q is equivalent to augmenting $S_{12} = f/g$ by the all-pass function Q_*/Q, i.e.

$$S_{12} \to S_{12} \cdot \frac{Q_*}{Q} \tag{6.28}$$

where, again, Q is chosen to contain all the zeros of ff_* in $\operatorname{Re} p < 0$ which are of odd multiplicity. Having thus augmented S_{12}, the strictly Hurwitz polynomial Q will appear automatically as a common factor in Z in Step (4). Therefore Step (5) can be eliminated from the procedure.

6.4 A PRELIMINARY SYNTHESIS STEP: THE BRUNE PREAMBLE

Starting from a p.r.f. $\hat{Z}(p)$, the synthesis procedure begins by extracting all $j\omega$-axis poles (including those at the origin and infinity) from $\hat{Z}(p)$ and its reciprocal $\hat{Y}(p)$. The standard methods of accomplishing this were discussed in Section 2.7. This process is called the *Brune preamble*, typical cycles of which are shown in Fig. 6.3. The remaining impedance is p.r. as shown in Corollary 2.7, and has equal degrees of its numerator and denominator. Such a function is clearly *minimum reactance* and *minimum susceptance* according to Definition 2.3.

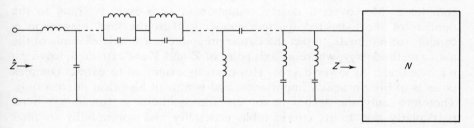

Fig. 6.3   The Brune preamble.

6.5 CASCADE SYNTHESIS

6.5.1 Outline of the technique

In the general theory of cascade synthesis to be discussed now, the starting-point is taken as the driving-point impedance $Z(p)$ which remains after the Brune preamble has been performed. $Z(p)$ is realized as a cascade of lossless reciprocal two-port subnetworks, terminated in a resistor as shown in Fig. 6.4. Each subnetwork in the cascade realizes a particular zero of transmission, i.e. a zero of the even-part of $Z(p)$.

Before proceeding with the explicit synthesis technique, it is worthwhile to explain the methodology of passive network synthesis in general, and cascade synthesis in particular. The fundamental idea behind any *valid*

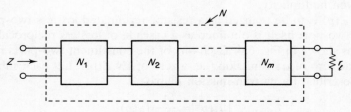

Fig. 6.4   Cascade synthesis.

synthesis technique is to extract from the given function $Z(p)$ a part that is recognizable as a physical subnetwork. Then, the remainder impedance is evaluated and the extraction process is repeated until the complete function is exhausted. However, in order for the process to be successful, the following conditions must be met.

(a) The extracted part must be realizable.
(b) The remainder must be positive real, i.e. the extraction does not destroy the p.r. character of the function.
(c) The remainder must be of lower degree than the original function.

Clearly, if the extraction cycle satisfies the above conditions, it can be repeated until a function of zero degree is reached, and the synthesis is completed. Moreover, if degree reduction in each cycle is equal to the number of the extracted reactive elements, then the final realization is canonic (or minimal). In fact the Brune preamble is a simple example of the above methodology, where $j\omega$-axis poles of Z and Y are extracted, leaving a p.r. remainder of lower degree. However, its extension to extract complex poles is of little practical importance, and is now of historical interest only. Therefore only the technique of cascade synthesis is considered here, particularly due to its considerable generality and applicability to filter synthesis.

6.5.2 The basic extraction procedure

Consider the minimum-reactance and minimum-susceptance function $Z(p)$ which remains after the Brune preamble (Fig. 6.3) has been performed. As shown in Section 5.7.2 the even-part of $Z(p)$ will still contain all the zeros of transmission originally present, except those which were $j\omega$-axis poles or zeros of $Z(p)$, since they have been removed in the Brune preamble. Moreover, the zeros of transmission must be of even multiplicity for a reciprocal realization. If some of these have odd multiplicity, the impedance $Z(p)$ is augmented as explained in Section 6.2, expressions (6.8) to (6.14). Alternatively if $Z(p)$ was derived from partial specification of the scattering matrix of the lossless two-port, as explained in Section 6.3.2, expressions (6.22) to (6.28), augmentation could be applied to the transfer function $S_{12}(p)$ as in (6.28), and the zeros of transmission are guaranteed to be of even multiplicity. So, we assume that the zeros of the even-part of $Z(p)$ are all of even multiplicity.

Now $Z(p)$ is to be realized as a resistor-terminated lossless two-port. The lossless two-port itself is obtained as a cascade of lossless reciprocal subnetworks, as shown in Fig. 6.4. Each one of the constituent two-ports realizes a particular zero of transmission (i.e. a zero of Ev $Z(p)$). Let each subnetwork N_i be described by its transmission matrix.

$$[T_i(p)] = \frac{1}{f_i(p)} [t_i(p)] \tag{6.29}$$

where $[t_i(p)]$ is the polynomial matrix whose properties were defined in Theorem 5.4,

$$[t_i(p)] = \begin{bmatrix} a_i(p) & b_i(p) \\ c_i(p) & d_i(p) \end{bmatrix} \tag{6.30}$$

The overall transmission matrix of the cascade N is given by

$$[T(p)] = \prod_{i=1}^{m} [T_i(p)]$$

$$= \prod_{i=1}^{m} \frac{1}{f_i(p)} [t_i(p)] \tag{6.31}$$

We also know from Theorem 5.5 in Section 5.7.2, that the zeros of transmission of a typical lossless reciprocal two-port described by $[t_i(p)]$ are those of the polynomial

$$\det [t_i(p)] = a_i d_i - b_i c_i \tag{6.32}$$

Therefore, the complete lossless two-port N made up of the cascade of N_i realizes all the zeros of transmission, which are the zeros of the polynomial

$$\det [t(p)] = \prod_{i=1}^{m} \det [t_i(p)]$$

$$= \prod_{i=1}^{m} f_i^2(p) \tag{6.33}$$

Now, write the given driving-point impedance as

$$Z(p) = \frac{m_1(p) + n_1(p)}{m_2(p) + n_2(p)} \tag{6.34}$$

with $m_{1,2}$ being even and $n_{1,2}$ odd polynomials. It was shown in Theorem 5.5 (Section 5.7.2) that the zeros of transmission are those of the even-part,

$$\text{Ev } Z(p) = \tfrac{1}{2}\{Z + Z_*\} \tag{6.35}$$

Henceforth, the argument p will be dropped for convenience. Using (6.34) and (6.35) the zeros of transmission are those of

$$u = m_1 m_2 - n_1 n_2 \tag{6.36}$$

Each zero of the above polynomial, is assumed to be of even multiplicity since augmentation of Z, if necessary, has been performed.

Since each subnetwork in the cascade of Fig. 6.4 is chosen such that it realizes particular zeros of transmission, a typical subnetwork N_i is called a *zero-producing section* or *zero-section* for short. We now examine the general procedure for extracting the zero-sections.

Let p_0 be a zero of transmission (i.e. of (6.36)), to be extracted by a section

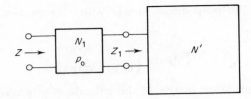

Fig. 6.5   Extraction of a zero section.

N_1 whose transmission matrix (still unknown) is

$$[T_1] = \frac{1}{f_1}[t_1]$$

$$= \frac{1}{f_1}\begin{bmatrix} a_1 & b_1 \\ c_1 & d_1 \end{bmatrix} \tag{6.37}$$

where, from (6.33) f_1^2 is the product of the factors of (6.36) corresponding to the zeros of transmission to be realized. Consider Fig. 6.5, and let Z_1 be the driving-point impedance of the remaining network after the extraction of the subnetwork with transmission matrix $[T_1]$ in (6.37). Thus, Z and Z_1 are related by

$$Z = \frac{a_1 Z_1 + b_1}{c_1 Z_1 + d_1} \tag{6.38}$$

or

$$Z_1 = \frac{d_1 Z - b_1}{a_1 - c_1 Z} \tag{6.39}$$

Now, it can be shown that if $[T_1]$ in (6.37) is chosen such that $\det [t_1] = f_1^2$ is a factor of u in (6.36), then the remaining impedance Z_1 is p.r. and of lower degree than Z. The reduction in degree is equal to the degree of f_1. The reader who is interested in the proof of this result may consult the paper by Youla[31] or that by Scanlan and Rhodes.[32]

Remember that we still do not know how to determine $[t_1]$ such that it satisfies the above criterion. This will be discussed shortly, but let us first obtain some simple expressions for Z and its derivative Z' ($\equiv dZ/dp$) evaluated at a zero of transmission p_0. These will be needed for the calculation of the transmission matrices of the subnetworks and for determining the element values of the realized network.

Let p_0 be a double-order zero of (6.36). Then

$$Z(p_0) = \frac{m_1(p_0) + n_1(p_0)}{m_2(p_0) + n_2(p_0)} \tag{6.40}$$

and

$$u(p_0) = (m_1 m_2 - n_1 n_2)_{p=p_0} = 0 \tag{6.41}$$

Combining (6.40) with (6.41) we obtain

$$Z(p_0) = \left(\frac{m_1}{n_2}\right)_{p=p_0} = \left(\frac{n_1}{m_2}\right)_{p=p_0} \tag{6.42}$$

Next consider the derivative of Z

$$Z' = \frac{(m_2+n_2)(m_1'+n_1') - (m_1+n_1)(m_2'+n_2')}{(m_2+n_2)^2} \tag{6.43}$$

where the prime denotes differentiation with respect to p. But p_0 is a *double-order* zero of u in (6.36). Hence $u' = du/dp$ contains the same zero p_0. Thus from (6.36) we have

$$(m_1m_2' + m_2m_1')_{p=p_0} = (n_1n_2' + n_2n_1')_{p=p_0} \tag{6.44}$$

Substitution from (6.42) and (6.44) into (6.43), gives

$$Z'(p_0) = \left(\frac{m_1}{n_2}\right)'_{p=p_0} = \left(\frac{n_1}{m_2}\right)'_{p=p_0} \tag{6.45}$$

6.5.3 The zero-sections

According to the location of the zeros of transmission in the p-plane, three types of zero-sections (i.e. two-port subnetworks) are needed. The first realizes a pair of conjugate $j\omega$-axis zeros, the second realizes a pair of zeros on the real axis, and the third realizes a complex conjugate quadruplet of zeros. In each case, remember that the zeros must be double-order zeros for a reciprocal lossless two-port. These zero-sections are now discussed and the expressions for the element values in each type are given. The detailed derivation of the element values is omitted for the sake of brevity, but this can be found in the references.[31,32]

THE BRUNE SECTION

This is a lossless reciprocal two-port subnetwork used to extract a pair of transmission-zeros on the $j\omega$-axis, i.e. at $p_0 = \pm j\omega_0$. Thus the Brune section realizes a pair of zeros corresponding to a factor of the form $(p^2 + \omega_0^2)^2$ in the even-part of the input impedance. The extraction results in degree reduction by two in the impedance. The transmission matrix of the Brune section is of the form

$$[T_B] = \frac{1}{\left(1 + \dfrac{p^2}{\omega_0^2}\right)} [t_B] \tag{6.46a}$$

with

$$[t_B] = \begin{bmatrix} 1 + k_1p^2 & k_2p \\ k_3p & 1 + k_4p^2 \end{bmatrix} \tag{6.46b}$$

where k_1, k_2, k_3, and k_4 are constants, to be determined from the function Z

and its derivative Z' evaluated at $p = j\omega_0$, as well as from the condition that $\det[t_B] = (1 + p^2/\omega_0^2)^2$. Let

$$Z(j\omega_0) = jX(\omega_0) = jx_0 \tag{6.47}$$

and

$$Z'(j\omega_0) = X'(\omega_0) = x_0' \tag{6.48}$$

where Z and Z' are obtained from (6.42) and (6.45). The constants in the transmission matrix (6.46) are then given by

$$k_1 = \frac{\omega_0 x_0' + x_0}{\omega_0^2(\omega_0 x_0' - x_0)} \tag{6.49}$$

$$k_3 = \frac{2}{\omega_0(\omega_0 x_0' - x_0)} \tag{6.50}$$

whereas k_2 and k_4 are dependent on k_1 and k_3 since they are related by

$$\det[t_B] = (1 + k_1 p^2)(1 + k_4 p^2) - k_2 k_3 p^2$$
$$= \left(1 + \frac{p^2}{\omega_0^2}\right)^2 \tag{6.51}$$

Hence,

$$k_4 = 1/k_1\omega_0^4 \tag{6.52}$$

$$k_2 = \frac{1}{k_3}\left(k_1 + k_4 - \frac{2}{\omega_0^2}\right) \tag{6.53}$$

Once these constants have been determined, $[T_B]$ in (6.46) is defined and can be realized by first finding the corresponding $[Z]$ or $[Y]$ of the two-port from (4.74) or (4.75), then the Brune section is obtained by the methods of Chapter 5 (Section 5.3 or 5.4). In fact, we gave Example 5.1 which is the synthesis of the impedance matrix of a Brune section.

The Brune section based on the synthesis of the impedance matrix is shown in Fig. 6.6, where the element values are given by

$$C = 2/\{\omega_0(\omega_0 x_0' - x_0)\} \tag{6.54a}$$

$$L_1 = (\omega_0 x_0' + x_0)/2\omega_0 \tag{6.54b}$$

$$L_2 = \frac{(\omega_0 x_0' - x_0)^2}{2\omega_0(\omega_0 x_0' + x_0)} \tag{6.54c}$$

$$M = (\omega_0 x_0' - x_0)/2\omega_0$$
$$= \sqrt{L_1 L_2} \tag{6.54d}$$

It remains to show, from the above expressions, that the element values C, L_1, and L_2 are guaranteed positive. First note that $Z(p_0)$ as obtained from (6.42) is a Foster function evaluated at $p_0 = j\omega_0$. This follows directly from Corollary 3.1. Thus by (3.15).

$$X'(\omega_0) > \frac{|X(\omega_0)|}{\omega_0}$$

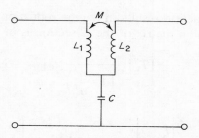

Fig. 6.6 Brune section $(M = \sqrt{L_1 L_2})$ or C-section $(M = -\sqrt{L_1 L_2})$ based on the synthesis of the impedance matrix.

or

$$\omega_0 x_0' - x_0 > 0 \qquad \text{for } \omega_0 > 0$$

It follows that all the expressions of $C, L_1,$ and L_2 in (6.54) are positive. Furthermore, the mutual inductance M is positive for the Brune section, with the coils being perfectly coupled.

It is also possible to obtain a realization of the Brune section based on its admittance matrix and the technique of Section 5.4. This is shown in Fig. 6.7, with the element values given by

$$C_1 = (\omega_0 x_0' + x_0)/2\omega_0 x_0^2 \tag{6.55a}$$

$$C_2 = -1/\omega_0 x_0 \tag{6.55b}$$

$$C_3 = \frac{\omega_0 x_0' + x_0}{\omega_0 x_0 (\omega_0 x_0' - x_0)} \tag{6.55c}$$

$$L = C_1/\omega_0^2 \tag{6.55d}$$

where C_1 and L are guaranteed positive, but either C_2 or C_3 is negative depending on the sign of x_0. Nevertheless this transformerless circuit is useful in filter design applications where the negative capacitor value may be absorbed by that of another cascaded section.

THE DARLINGTON C-SECTION

This realizes a pair of transmission-zeros on the real axis, i.e. $p_0 = \pm \sigma_0$. These correspond to a factor of the form $(p^2 - \sigma_0^2)^2$ in the even-part of the

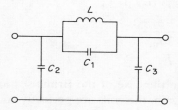

Fig. 6.7 Brune section based on the synthesis of the admittance matrix.

impedance. The extraction results in degree reduction by two in the impedance. The transmission matrix of the C-section is of the form

$$[T_C] = \frac{1}{\left(1 - \dfrac{p^2}{\sigma_0^2}\right)} [t_C] \tag{6.56a}$$

with

$$[t_C] = \begin{bmatrix} 1 + k_1 p^2 & k_2 p \\ k_3 p & 1 + k_4 p^2 \end{bmatrix} \tag{6.56b}$$

where k_1, k_2, k_3, and k_4 are real constants which can be determined in a
manner similar to that employed for the Brune section. Thus

$$k_1 = \frac{\sigma_0^2(\sigma_0 z_0' + z_0)}{z_0 - \sigma_0 z_0'} \tag{6.57}$$

$$k_3 = \frac{2\sigma_0}{(z_0 - \sigma_0 z_0')} \tag{6.58}$$

where

$$z_0 = Z(\sigma_0), \quad z_0' = Z'(\sigma_0) \tag{6.59}$$

which can be calculated using (6.42) and (6.45). The other constants are not
independent and can be obtained from (6.57), (6.58), and the condition,

$$\det [t_C] = (1 + k_1 p^2)(1 + k_4 p^2) - k_2 k_3 p^2$$

$$= \left(1 - \frac{p^2}{\sigma_0^2}\right)^2 \tag{6.60}$$

Again, once the constants in $[t_C]$ have been determined, the impedance or
admittance matrix can be obtained using (4.74), (4.75), then the synthesis of
the section is accomplished using the techniques of Section 5.3 or 5.4. In
fact, Example 5.2 given in Section 5.3 dealt with the synthesis of the
impedance matrix of a C-section.

The C-section based on the synthesis of the impedance matrix has the
same form as the Brune section of Fig. 6.6. However, the mutual inductance
is now negative, and the element values are given by

$$C = 2/\{\sigma_0(z_0 - \sigma_0 z_0')\} \tag{6.61a}$$

$$L_1 = (z_0 + \sigma_0 z_0')/2\sigma_0 \tag{6.61b}$$

$$L_2 = \frac{(z_0 - \sigma_0 z_0')^2}{2\sigma_0(z_0 + \sigma_0 z_0')} \tag{6.61c}$$

$$M = -\frac{(z_0 - \sigma_0 z_0')}{2\sigma_0}$$

$$= -\sqrt{L_1 L_2} \tag{6.61d}$$

Similar reasoning to that employed in the case of the Brune section, shows
that C, L_1, and L_2 are all positive. Again, the coils are perfectly coupled but
the mutual inductance M is negative in this case.

The observant reader must have noticed that expressions (6.61) for the element values of the C-section can be obtained from those in (6.54) for the Brune section, if ω_0 is replaced by $j\sigma_0$, x_0 by jz_0, and x'_0 by z'_0. Therefore, if the admittance matrix was used instead of the impedance matrix to realize the C-section, the same circuit in Fig. 6.7 would result, with the element values given by (6.55) with ω_0 replaced by $j\sigma_0$, x_0 by jz_0, and x'_0 by z'_0. But this would force C_1 to be negative and cannot be absorbed by other sections. Therefore, this realization of the C-section from the admittance matrix is of no practical use.

THE DARLINGTON D-SECTION

This realizes a complex conjugate quadruplet of transmission zeros at $p_0 = \pm(\sigma_0 \pm j\omega_0)$, corresponding to a factor of the form $\{p^4 + 2(\omega_0^2 - \sigma_0^2)p^2 + |p_0|^4\}^2$ in the even-part of the impedance. The extraction of a D-section results in degree reduction by four in the impedance. The transmission matrix of the D-section is of the form

$$[T_D] = [t_D]/\{p^4 + 2(\omega_0^2 - \sigma_0^2)p^2 + |p_0|^4\} \tag{6.62a}$$

where

$$[t_D] = \begin{bmatrix} |p_0|^4 + k_1 p^2 + k_2 p^4 & p(k_3 + k_4 p^2) \\ p(k_5 + k_6 p^2) & |p_0|^4 + k_7 p^2 + k_8 p^4 \end{bmatrix} \tag{6.62b}$$

Evaluation of the constants k_1 to k_8 is given in the references,[31,32] and involves much heavier algebra than the previous sections. To simplify the expressions for the element values, let

$$\begin{aligned} Z(\sigma_0 + j\omega_0) &= r + jx \\ Z'(\sigma_0 + j\omega_0) &= r' + jx' \end{aligned} \tag{6.63}$$

and define the quantities,

$$J_1 = x\sigma_0^3 - r'\sigma_0\omega_0(\sigma_0^2 + \omega_0^2) + r\omega_0^3 \tag{6.64a}$$

$$J_2 = -x\sigma_0^3 + r'\sigma_0\omega_0(\sigma_0^2 - \omega_0^2) - 2x'\sigma_0^2\omega_0^2 + r\omega_0^3 \tag{6.64b}$$

$$J_3 = x'(\sigma_0^2 + \omega_0^2) + 3\omega_0 r - 3\sigma_0 x \tag{6.64c}$$

$$J_4 = -x'(\sigma_0^2 - \omega_0^2) - 2\sigma_0\omega_0 r' + \omega_0 r + \sigma_0 x \tag{6.64d}$$

$$H_1 = J_2\{x\sigma_0(3\omega_0^2 - \sigma_0^2) + r\omega_0(3\sigma_0^2 - \omega_0^2)\}$$
$$\qquad + J_1(\sigma_0^2 + \omega_0^2)(r\omega_0 - x\sigma_0) \tag{6.64e}$$

$$H_2 = 1/(J_2 J_3 - J_1 J_4) \tag{6.64f}$$

With the above abbreviations, we have

$$k_1 = 2(\omega_0^2 - \sigma_0^2) + 4H_1 H_2 \tag{6.65a}$$

$$k_2 = 1 + 4H_2\{J_1(x\sigma_0 + r\omega_0) + J_2(x\sigma_0 - r\omega_0)\} \tag{6.65b}$$

$$k_5 = 8\sigma_0\omega_0(\sigma_0^2 + \omega_0^2)J_2 H_2 \tag{6.65c}$$

$$k_6 = 8\sigma_0\omega_0 J_1 H_2 \tag{6.65d}$$

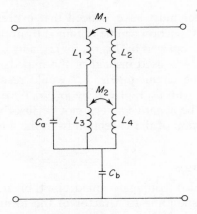

Fig. 6.8 D-section based on the
synthesis of the impedance matrix.

The other constants in $[t_D]$ are dependent on the above ones, and can be evaluated using (6.65) and the relation

$$\det[t_D] = \{p^4 + 2(\omega_0^2 - \sigma_0^2)p^2 + |p_0|^4\}^2 \tag{6.66}$$

The realization of the D-section is then accomplished by first evaluating its impedance or admittance matrix from (6.62) and (4.74) to (4.75), then using the techniques of Section 5.3 or 5.4. A realization based on the impedance matrix is shown in Fig. 6.8 with the element values given by

$$M_1 = 1/k_6 \tag{6.67a}$$

$$L_1 = k_2/k_6 \tag{6.67b}$$

$$L_2 = M_1^2/L_1 \tag{6.67c}$$

$$M_2 = \frac{-1}{k_5^2 k_6}\{k_5^2 - 2(\omega_0^2 - \sigma_0^2)k_5 k_6 + |p_0|^4 k_6^2\} \tag{6.67d}$$

$$L_3 = \frac{1}{k_5^2 k_6}\{k_1 k_5 k_6 - k_2 k_5^2 - |p_0|^4 k_6^2\} \tag{6.67e}$$

$$L_4 = M_2^2/L_3 \tag{6.67f}$$

$$C_a = k_6/k_5 L_3 \tag{6.67g}$$

$$C_b = k_5/|p_0|^4 \tag{6.67h}$$

All the capacitor and inductor values can be proved positive. We also note from the above expressions that both pairs of coils M_1 and M_2 are perfectly-coupled.

An alternative realization of the D-section may be obtained from its admittance matrix, and can be found in the references.[32]

6.5.4 The complete synthesis—recapitulation

Having discussed the three types of zero-sections, it only remains to explain the details of how the complete synthesis of the impedance is achieved. It is also worthwhile to summarize the entire procedure.

(i) The starting function $Z(p)$ is assumed minimum-reactance and minimum-susceptance, since the Brune preamble in Section 6.4 has been performed, if necessary. This $Z(p)$ is to be realized as a resistor-terminated cascade of lossless reciprocal subnetworks as shown in Fig. 6.4.

(ii) The even-part of $Z(p)$ is evaluated, and its zeros are the zeros of transmission of the lossless two-port N of Fig. 6.4. These zeros are assumed to be of even multiplicity, for a reciprocal realization, since augmenting $Z(p)$ as explained in (6.8) is always possible.

(iii) Each (double-order) zero of transmission is realized by a Brune, C, or D-type section depending on the location of the zero in the p-plane.

(iv) The process begins by extracting any zero of transmission. The transmission matrix of the extracted zero-section is calculated using the appropriate expressions: (6.46), (6.56), or (6.62). The element values of the section are obtained from (6.54), (6.61), or (6.67).

(v) The entries of the transmission matrix of the extracted zero-section are then substituted in (6.39) to obtain the remainder impedance Z_1 of the network after the extraction of this particular section. This is guaranteed p.r., and of lower degree than the original Z.

(vi) Another zero-section is extracted from Z_1 as calculated in (v). This zero-section realizes another zero of transmission, and the remaining impedance is calculated as in (v).

(vii) The extraction of zero-sections in cascade is repeated until all the zeros of transmission are realized, and we reach an impedance whose value equals the terminating resistor.

Example 6.1 Realize the following p.r. impedance as a resistor-terminated lossless reciprocal two-port

$$Z(p) = \frac{8p^4 + 33p^3 + 18p^2 + 6p + 2}{2p^4 + 4p^3 + 13p^2 + 5p + 2}$$

Solution First note that the given impedance is devoid of $j\omega$-axis poles and zeros including the origin and infinity. Hence Z is minimum-reactance and minimum-susceptance. Next write

$$Z(p) = \frac{(8p^4 + 18p^2 + 2) + (33p^3 + 6p)}{(2p^4 + 13p^2 + 2) + (4p^3 + 5p)}$$

$$= \frac{m_1 + n_1}{m_2 + n_2}$$

According to (6.36) the zeros of Ev Z are the zeros of transmission, and are the zeros of the polynomial

$$m_1 m_2 - n_1 n_2 = (4p^2 + 1)^2(p^4 + 4)$$

The $j\omega$-axis zeros at $p = \pm j\frac{1}{2}$ are double, and are realizable by a Brune section. On the other hand, the complex conjugate quadruplet at $p = \pm(\pm j1)$ corresponds to the simple factor $(p^2 + 4)$. Thus, Z must be augmented to supply another factor $(p^4 + 4)$ in the even-part. This augmentation, however, can be postponed until after the extraction of the Brune section realizing the factor $(4p^2 + 1)^2$. The remaining impedance can then be augmented and a D-section extracted. Thus, for the Brune section, using (6.42) and (6.45) we have

$$Z(j\tfrac{1}{2}) = \left(\frac{m_1}{n_2}\right)_{p=j\frac{1}{2}} = j1$$

or

$$x_0 = 1$$

and

$$Z'(j\tfrac{1}{2}) = 6$$

or

$$x_0' = 6$$

Substituting in (6.54) we obtain the element values by reference to Fig. 6.6 as:

$$L_1 = 4\,\text{H}, \quad L_2 = 1\,\text{H}, \quad M = 2\,\text{H}, \quad C = 2\,\text{F}$$

The remaining impedance after the above extraction, must now be calculated. Using (6.49), (6.50), (6.52), and (6.53) we first evaluate the constants k_1 to k_4 and hence the transmission matrix of the extracted Brune section. Its entries are then used in (6.39) to calculate the remainder Z_1. After the cancellation of a common factor $(4p^2 + 1)^2$, this gives

$$Z_1(p) = \frac{p^2 + 4p + 2}{p^2 + p + 2}$$

The numerator of Ev Z_1 is $(p^4 + 4)$ as expected. Z_1 must now be augmented as

$$Z_1 = \frac{p^2 + 4p + 2}{p^2 + p + 2} \cdot \frac{Q(p)}{Q(p)}$$

where $Q(p)$ contains all the left half-plane zeros of $(p^4 + 4)$. Thus

$$Q(p) = \{p + (1 + j1)\}\{p + (1 - j1)\}$$
$$= p^2 + 2p + 2$$

Therefore the augmented Z_1 becomes

$$Z_1(p) = \frac{p^4 + 6p^3 + 12p^2 + 12p + 4}{p^4 + 3p^3 + 6p^2 + 6p + 4}$$

$$= \frac{(p^4 + 12p^2 + 4) + (6p^3 + 12p)}{(p^4 + 6p^2 + 4) + (3p^3 + 6p)}$$

$$= \frac{M_1 + N_1}{M_2 + N_2}$$

and the numerator of its even-part is

$$M_1 M_2 - N_1 N_2 = (p^4 + 4)^2$$

Hence the complex zero is double-order as required. This is realized by a D-section. From (6.42) and (6.45) we have

$$Z_1(1 + j1) = 2$$

$$Z_1'(1 + j1) = -\tfrac{1}{3} - j\tfrac{1}{3}$$

Substitution of the above values in (6.63) to (6.64) gives

$$J_1 = 8/3, \quad J_2 = 8/3, \quad J_3 = 16/3, \quad J_4 = 8/3$$

The element values are then calculated from (6.65), (6.67) for the network of Fig. 6.8. Finally, the remainder impedance after the extraction of the D-section is obtained from (6.39) and is seen to be a $1\,\Omega$-resistor. The complete realized network is shown in Fig. 6.9.

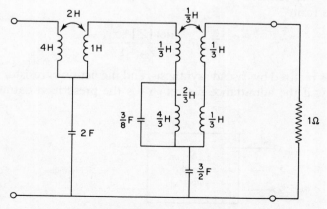

Fig. 6.9   Network of Example 6.1.

6.6 CASCADE SYNTHESIS OF [*T*], [*Z*], AND [*Y*] OF A LOSSLESS TWO-PORT

The theory of cascade synthesis discussed in the previous section solves two problems at once. The first is that specifically considered so far, namely: the synthesis of an arbitrary p.r. impedance. But it is evident that the Darlington technique reduces the driving-point synthesis of an RLC impedance to the synthesis of a lossless reciprocal two-port (terminated in a resistor). Consequently, we actually know the general cascade synthesis technique of a lossless reciprocal two-port: terminated or unterminated.

Consider an unterminated lossless reciprocal two-port *N* as shown in Fig. 6.10. Let the prescribed datum be the polynomial transmission matrix [*t*] which satisfies the conditions of Theorem 5.4, Corollary 5.3 for a lossless reciprocal two-port. Then if we *assume* that the two-port is terminated in a *fictitious* 1 Ω-resistor, the input impedance of the one-port thus formed is

$$Z_{in}(p) = \frac{a(p) + b(p)}{c(p) + d(p)} \tag{6.68}$$

where *a*, *b*, *c*, and *d* are the entries of the polynomial transmission matrix defined in Section 5.6. Hence, the entire technique and expressions for element values, etc., of cascade synthesis can be applied to Z_{in}. Of course the 1 Ω resistor at the output is omitted since it is only a means for allowing the use of the formulae and concepts discussed earlier. If it so happens that the prescribed [*t*] satisfies the conditions of Theorem 5.4 for a lossless *non-reciprocal* network, then we still form Z_{in} in (6.68) and augment as before for a reciprocal realization.

Next, consider the case where the prescribed datum is the impedance matrix [*Z*] of the lossless two-port *N*. Suppose that [*Z*] is a Foster matrix satisfying the conditions of Theorem 5.1 or Corollary 5.1. Again we assume a fictitious 1 Ω resistor as a load (see Fig. 6.10) and calculate the input impedance from

$$Z_{in}(p) = \frac{\det[Z] + z_{11}}{z_{22} + 1} \tag{6.69}$$

Then Z_{in} is realized by cascade synthesis, and the fictitious resistor is omitted.

Similarly, if the admittance matrix [*Y*] is the prescribed datum, and is a

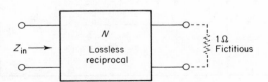

Fig. 6.10   Pertinent to cascade synthesis of a lossless reciprocal two-port from its [*Z*], [*Y*], or [*T*].

Foster matrix, then we form the input impedance of a fictitious-1 Ω-terminated one-port as

$$Z_{in} = \frac{y_{22}+1}{\det[Y]+y_{11}} \qquad (6.70)$$

which is realized by cascade synthesis and the fictitious resistor is omitted.

6.7 A NOTE ON ALL-PASS TWO-PORTS

The transmission coefficient of an all-pass two-port is of the form

$$S_{21}(p) = \pm\frac{g_*}{g} \qquad (6.71)$$

so that $|S_{21}(j\omega)| = 1$ at all frequencies. The corresponding input impedance of the $1\,\Omega$-terminated lossless two-port is given by

$$Z(p) = \frac{g}{g} = 1 \qquad (6.72)$$

and the all-pass lossless two-port can be realized using cascade synthesis by realizing the zeros of gg_*. This requires C and/or D-sections but never Brune sections, since g is strictly Hurwitz. The same formulae for the C and D sections given in Section 6.5.3 can be used directly to obtain the element values and realize the all-pass two-port. However, in using (6.42) and (6.45) the simplifications

$$Z = 1, \quad Z' = 0 \qquad (6.73)$$

occur. All-pass networks are of importance since they can be used as phase equalizers without affecting the amplitude response of the filter. This will be explained in a later chapter.

6.8 CONCLUSION

In many ways, the present chapter represents the culmination of the preceding development. We have established the sufficiency of the p.r. condition for realizability of a given function $Z(p)$ as the driving-point impedance of a resistor-terminated lossless reciprocal two-port. At once, this solves the synthesis of lossless reciprocal two-ports and establishes the link with filter synthesis. The most general realization requires perfectly-coupled coils (or ideal transformers). The problem of transformerless synthesis of an arbitrary p.r.f. has always attracted the attention of circuit theorists. Bott and Duffin[33] showed that it is possible to realize a p.r.f. as an RLC one-port without transformers. However, the required number of elements is excessively large and requires a number of resistors to be distributed throughout the network. These factors render the Bott–Duffin technique of little practical use, although it is of theoretical importance. The most recent contribution on the transformerless synthesis of an arbitrary p.r.f. is due to Fialkow.[34,35]

PROBLEMS

6.1 Consider the p.r. impedance

$$Z(p) = \frac{p^4 + 2p^3 + 6p^2 + 2p + 4}{p^3 + p^2 + 4p}$$

(a) Perform the Brune preamble on $Z(p)$, i.e. extract all $j\omega$-axis poles of Z and Y, including those at $p = 0$, $p = \infty$.
(b) Evaluate the even-part of the remainder impedance.
(c) Use cascade synthesis to realize the remainder, hence the entire impedance.

6.2 Realize the following minimum reactance and minimum susceptance p.r.f. using the technique of cascade synthesis.

$$Z(p) = \frac{p^4 + 4p^3 + 3p^2 + 4p + 1}{p^4 + p^3 + 3p^2 + p + 1}$$

6.3 Given the polynomial matrix

$$[t(p)] = \begin{bmatrix} 9p^2 + 1 & 2p^3 + 3p \\ 18p^3 + 3p & 4p^4 + 8p^2 + 1 \end{bmatrix}$$

verify that $[t(p)]$ is realizable as the polynomial transmission matrix of a lossless reciprocal two-port. Find a realization by cascade synthesis.

6.4 A lossless two-port has the transmission coefficient

$$S_{21}(p) = \frac{3}{p^2 + 3p + 3}$$

referred to $1\,\Omega$ terminations. Calculate the input impedance of the $1\,\Omega$-terminated lossless two-port, and hence find a realization.

6.5 Realize the following Foster matrices by cascade synthesis

(a)

$$[Z(p)] = \begin{bmatrix} \dfrac{2 + p^2}{p} & \dfrac{2 - 2p^2}{p} \\ \dfrac{2 - 2p^2}{p} & \dfrac{2 + 4p^2}{p} \end{bmatrix}$$

(b)

$$[Z(p)] = \begin{bmatrix} \dfrac{p^4 + 3p^2 + 1}{p^3 + p} & \dfrac{2p^4 + 2p^2 + 1}{p^3 + p} \\ \dfrac{2p^4 + 2p^2 + 1}{p^3 + p} & \dfrac{4p^4 + 6p^2 + 1}{p^3 + p} \end{bmatrix}$$

(c)
$$[Y(p)] = \begin{bmatrix} \dfrac{p^2+1}{p} & -\dfrac{p^2+2}{p} \\ -\dfrac{p^2+2}{p} & \dfrac{p^2+4}{p} \end{bmatrix}$$

(d)
$$[Y(p)] = \begin{bmatrix} \dfrac{p^4+4p^2+1}{p^3+p} & -\dfrac{p^4+p^2+2}{p^3+p} \\ -\dfrac{p^4+p^2+2}{p^3+p} & \dfrac{p^4+7p^2+4}{p^3+p} \end{bmatrix}$$

6.6 Consider the all-pass transmission coefficient, referred to $1\,\Omega$-terminations,

$$S_{21}(p) = \frac{4-6p+4p^2-p^3}{4+6p+4p^2+p^3}$$

Find a reciprocal realization by cascade synthesis.

6.7 Consider the singly terminated two-port N shown in Fig. P6.7. Define the transfer impedance as

$$Z_{21}(p) = \frac{V_2(p)}{I_1(p)}$$

Derive the necessary and sufficient conditions under which Z_{21} is realizable as the transfer impedance of a lossless reciprocal two-port, hence develop a synthesis technique (*Hint*: express Z_{21} in terms of the transmission parameters of N and compare with the conditions of Theorem 5.4 and Corollary 5.3.)

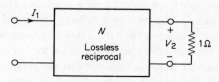

Fig. P6.7   Problem 6.7.

6.8 Using the results of Problem 6.7, realize the transfer impedance

$$Z_{21}(p) = \frac{2p^2+1}{4p^2+p+1}$$

as that of a $1\,\Omega$-terminated lossless reciprocal two-port.

Chapter 7
Synthesis of Special Configurations

7.1 INTRODUCTION

We shall see in a later part of the book, that most passive filters are designed in ladder form as shown in Fig. 7.1. The component series and shunt arms are one-port reactances, i.e. Z_i $(i = 0, 1, 2, \ldots, n)$ are Foster functions. In addition to the desirable running ground, the lossless ladder has the very important property that all its zeros of transmission are included in the open-circuit frequencies of the series arms, and the short-circuit frequencies of the shunt arms. They are either poles of $Z_1, Z_3, \ldots, Z_{n-1}$ or zeros of Z_0, $Z_2, Z_4, \ldots, Z_n$.

Furthermore, many active filters are modelled on the passive lossless ladder, with the objectives of imitating the low sensitivity properties of the ladder as well as exploiting the wealth of material available for the synthesis of such structures.

As shown in Chapter 6, an arbitrary p.r. impedance is realizable as a lossless reciprocal two-port terminated in a resistor. However, the realization requires, in general, the use of perfectly coupled coils and/or ideal transformers. We have also seen that the resulting network is defined by the location of the zeros of transmission. Thus, in order for the input impedance of the resistor-terminated lossless two-port to be realizable in a prescribed structure, particularly the ladder configuration, it must satisfy some *additional constraints as well as being positive real*. Consideration in this chapter is given to some special types of p.r. impedances or bounded real transmission coefficients $S_{21}(p)$ which are realizable in special forms, in particular the ladder structure.

7.2 SIMPLE LOW-PASS LOSSLESS LADDER

This is the structure of Fig. 7.2 with series inductors and shunt capacitors. Depending on whether the starting point is $Z(p)$, $S_{21}(p)$, or $[t(p)]$, we have three possible ways of stating the realizability conditions for this type of low-pass ladder.

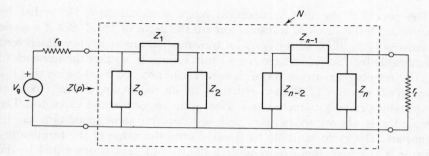

Fig. 7.1 General form of lossless ladder.

7.2.1 The conditions in terms of the input impedance

Let $Z(p)$ be a p.r.f. written as

$$Z(p) = \frac{N}{D}$$

$$= \frac{m_1 + n_1}{m_2 + n_2} \tag{7.1}$$

with the usual notation of $m_{1,2}$ being even and $n_{1,2}$ being odd. Suppose that the even-part of $Z(p)$

$$\text{Ev } Z = \frac{1}{2} \frac{N_* D + N D_*}{D D_*}$$

$$= \frac{m_1 m_2 - n_1 n_2}{m_2^2 - n_2^2} \tag{7.2}$$

is such that

$$\text{Ev } Z = \frac{K}{m_2^2 - n_2^2} \tag{7.3}$$

where K is a constant. Then $Z(p)$ is realizable as the driving-point impedance of a resistor-terminated lossless ladder two-port N of the type shown in Fig. 7.2.

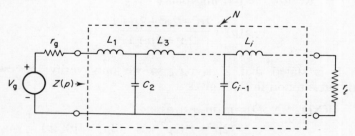

Fig. 7.2 Simple low-pass lossless ladder.

The proof of the above condition is quite elementary. Note that by Theorem 5.5 the zeros of transmission are the zeros of Ev Z. But if a given Z satisfies (7.3), then all the zeros of transmission are at infinity. Each zero is realizable by a series inductor or a shunt capacitor, so that the network of Fig. 7.2 results. A rigorous proof, however, can proceed by showing that if a given p.r.f. satisfies (7.3) then either Z or its reciprocal Y has a pole at $p = \infty$, which can be extracted as a series inductor or a shunt capacitor. The remainder is shown to be p.r. and satisfies the same condition on its even-part. The cycle can then be iterated until the value of the terminating resistor is reached; this occurs after a number of extractions equal to the degree of the given impedance. The reader who has gone so far in the book should be able to fill in the details of the proof. From the given interpretation, the extraction procedure consists in the successive extraction of poles at $p = \infty$ from Z and the inverted remainders. Hence the algebraic method of continued fractions can be employed to realize the entire impedance. Thus

$$Z(p) = \frac{N}{D}$$

$$= \alpha_1 p + \cfrac{1}{\alpha_2 p + \cfrac{1}{\alpha_3 p + {}_{\displaystyle \cdot_{\displaystyle \cdot_{\displaystyle \cdot}}}}}$$

$$\alpha_n p + \frac{1}{r_e} \text{ (or } r_e) \tag{7.4}$$

where the α_i's are the element values in Fig. 7.2, and r_e is the value of the terminating resistor. Note that in (7.4), it is assumed that Z has a pole at $p = \infty$. If, instead, Y has a pole at $p = \infty$, the first step is inversion, and the process begins from the admittance. In this case the first element is a shunt capacitor. Also, depending on whether the last element is a series inductor or a shunt capacitor, the final remainder is r_e or $1/r_e$ respectively.

Example 7.1 Realize the p.r. impedance

$$z(p) = \frac{2p^3 + 2p^2 + 2p + 1}{2p^2 + 2p + 1}$$

Solution It is stated that Z is p.r., so we only verify the even-part condition. Substitution in (7.2) gives

$$\tfrac{1}{2}(ND_* + N_*D) = m_1 m_2 - n_1 n_2$$
$$= (2p^2 + 1)(2p^2 + 1) - (2p^3 + 2p)(2p)$$
$$= 1$$

Thus Z satisfies the even-part constraint in (7.3) with $K = 1$. It is, therefore realizable as in Fig. 7.2. Since Z has a pole at $p = \infty$, the continued fraction is applied directly.

$$
2p^2+2p+1 \overline{\left) \begin{array}{l} 2p^3+2p^2+2p+1 \\ \underline{2p^3+2p^2+p} \end{array} \right.} \, \genfrac{}{}{0pt}{}{p}{}
$$

$$
\begin{array}{c|l}
 & 2p \\
\hline
p+1 & 2p^2+2p+1 \\
 & \underline{2p^2+2p} \\
\hline
 & 1 \quad \begin{array}{l|l} & p \\ \hline p+1 & \\ \underline{p} & \end{array} \\
 & \qquad\qquad 1
\end{array}
$$

Therefore, by reference to Fig. 7.2, the element values in the ladder, with $n = 3$, are: $L_1 = 1\,\text{H}$, $C_2 = 2\,\text{F}$, $L_3 = 1\,\text{H}$, $r_\ell = 1\,\Omega$.

7.2.2 The conditions in terms of the transmission coefficient

Let $S_{21}(p)$ be a *bounded real* function of the *specific* form

$$S_{21}(p) = \frac{1}{g_n(p)} \tag{7.5a}$$

or

$$|S_{21}(j\omega)|^2 = \frac{1}{|g_n(j\omega)|^2} \tag{7.5b}$$

Then, S_{21} is realizable as the transmission coefficient, referred to real terminations, of a lossless ladder two-port N with series inductors and shunt capacitors as shown in Fig. 7.2. The number of reactive elements is equal to n, the degree of S_{21}.

The above conditions follow directly from the interpretation of the zeros of S_{21} as the zeros of transmission, which in the case of (7.5) are all at $p = \infty$. The synthesis is performed by first deriving the input impedance, using Steps (1) to (4) in Section 6.3.2, which is then realized by the same method of Section 7.2.1, i.e. using (7.4).

7.2.3 The conditions in terms of the transmission matrix

Let $[t(p)]$ be a 2×2 polynomial matrix, satisfying the conditions of Theorem 5.4, Corollary 5.3 for the polynomial transmission matrix of a lossless reciprocal two-port, but with the added constraint,

$$\det [t(p)] \equiv f^2(p)$$

$$= \text{a constant} \tag{7.6}$$

Then $[t(p)]$ is realizable as lossless ladder N of the form shown in Fig. 7.2. Again, since det $[t]$ is a constant, it follows from Theorem 5.5 that the lossless two-port has no finite zeros of transmission. Therefore, they are all at $p = \infty$. The synthesis of $[t(p)]$ can be performed by first constructing the input impedance of the one-port obtained from N by closing its output on a fictitious 1Ω resistor according to expression (6.68). Then the impedance is realized as explained in Section 7.2.1 by expanding as in (7.4). The fictitious resistor can then be discarded.

Finally, we note that the conditions given here on $[t]$ and those given before in Sections 7.2.1 and 7.2.2 on Z and S_{21}, are all equivalent. They impose the requirement that all the transmission zeros of the lossless two-port must lie at $p = \infty$ for a realization as a low-pass ladder with series inductors and shunt capacitors.

7.3 SIMPLE HIGH-PASS LOSSLESS LADDER

This is the structure of Fig. 7.3 with series capacitors and shunt inductors. The realizability conditions for this form can be easily obtained from those in the previous section if we transform the point $p = \infty$ to the point $p = 0$ by letting $p \rightarrow 1/p$ in all the results of Section 7.2. The lossless two-port now has all its transmission zeros at $p = 0$. Consequently we have the following results:

(i) A p.r.f. $Z(p) = N/D$ is realizable as the driving-point impedance of a resistor-terminated lossless ladder two-port with series capacitors and shunt inductors if and only if

$$\text{Ev } Z = \frac{Kp^{2n}}{D_n D_{n*}} \qquad (7.7)$$

where K is a constant and n is the degree of Z.

(ii) A bounded real function $S_{21}(p)$ is realizable as the transmission coefficient of a lossless ladder two-port N of the form shown in Fig. 7.3 if and only if it is of the form

$$S_{21}(p) = \frac{p^n}{g_n(p)} \qquad (7.8)$$

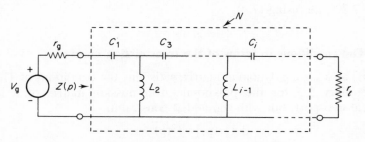

Fig. 7.3   Simple high-pass lossless ladder.

(iii) The 2×2 polynomial matrix $[t(p)]$ is realizable as the polynomial transmission matrix of the high-pass lossless ladder two-port N shown in Fig. 7.3 if and only if $[t(p)]$ satisfies the conditions of Theorem 5.4 with

$$\det [t(p)] \equiv f^2 = Kp^{2n} \tag{7.9}$$

with K as a positive constant.

The synthesis of the impedance given in (i), or obtained from specifying S_{21} or $[t]$ in (ii) or (iii) is accomplished by extracting poles at $p = 0$ from Z and the subsequently inverted remainders. The numerator and denominator of Z are arranged in ascending powers of p, and the starting function Z or Y must have a pole at $p = 0$. Then the continued fraction expansion (around $p = 0$) of the function is found as

$$Z(p) = \cfrac{1}{\beta_1 p} + \cfrac{1}{\cfrac{1}{\beta_2 p} + 1}$$

$$\cfrac{1}{\beta_n p} + r_e\left(\text{or } \frac{1}{r_e}\right) \tag{7.10}$$

in which β_i's are the element values, n is the degree of the function, and r_e is the value of the terminating resistor. Again, if Y (not Z) has the pole at $p = 0$, the expansion starts from Y and the first element is a shunt inductor.

7.4   A LOSSLESS ONE-PORT IN SERIES WITH A RESISTOR

Consider the special case of a p.r. impedance $Z(p)$ which has all its *poles* on the $j\omega$-axis. Note that the zeros may be anywhere in the left half-plane. Then $Z(p)$ may be written as

$$Z(p) = \frac{N(p)}{D(p)} \tag{7.11}$$

where $D(p)$ is either even or odd, but N is a general Hurwitz polynomial. In this case, extraction of all the poles of Z according to Corollary 2.7, leaves a remainder equal to the terminating resistor. Therefore, the impedance is realized completely by the technique of Section 2.7, and in (2.57) $Z_1 = a$ constant. By reference to Fig. 2.8 the network N_1 is a simple resistor. Also, using (5.66) the condition that Z has only $j\omega$-axis poles is equivalent to

$$\text{Ev } Z = a \text{ constant} \tag{7.12}$$

Thus, $Z(p)$ is realizable as a lossless one-port in series with a resistor, as shown in Fig. 7.4.

Similarly, a p.r. admittance $Y(p)$ which has only $j\omega$-axis poles is realizable as a lossless one-port in parallel with a resistor.

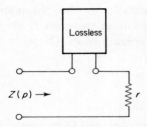

Fig. 7.4 A lossless one-port in series with a resistor.

7.5 MID-SHUNT AND MID-SERIES LOW-PASS LOSSLESS LADDERS

These are shown in Fig. 7.5, and are of considerable importance in filter synthesis. It turns out, as we shall see in a later chapter, that the optimum solution to the amplitude approximation problem requires the use of transfer functions $S_{21}(p)$ possessing finite zeros of trransmission on the $j\omega$-axis. But we have seen in the general theory of cascade synthesis discussed in Section 6.5 that the synthesis of such functions requires, in general, perfectly coupled coils (or ideal transformers). It follows that if we wish to employ a ladder network of the type shown in Fig. 7.5 to realize finite $j\omega$-axis zeros of transmission, additional constraints must be imposed on the driving-point impedance. These conditions were given by Fujisawa,[36] and are now stated.

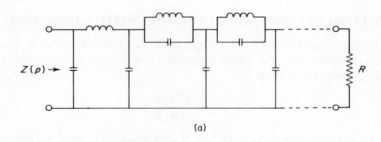

(a)

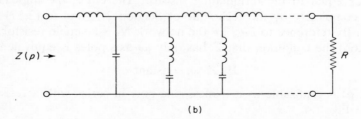

(b)

Fig. 7.5 General forms of low-pass mid-shunt and mid-series lossless ladders. (a) Mid-shunt, (b) mid-series.

Theorem 7.1 (Fujisawa) *Let the given impedance be written as*

$$Z(p) = \frac{N(p)}{D(p)}$$

$$= \frac{m_1 + n_1}{m_2 + n_2} \tag{7.13}$$

where N and D are relatively prime, with the usual notation of $m_{1,2}$ being even and $n_{1,2}$ being odd. Then $Z(p)$ represents the driving-point impedance of a low-pass mid-shunt lossless ladder terminated in a resistor R as shown in Fig. 7.5(a) if and only if

(i) $Z(p)$ *is positive real.*

(ii) $Z(0) = R.$ $\hspace{4cm}$ (7.14)

(iii) *At $p = \infty$, $Z(p)$ has either a pole or a zero.*

(iv)

$$\tfrac{1}{2}(ND_* + N_*D) = m_1 m_2 - n_1 n_2$$

$$= K \prod_{i=1}^{l} (p^2 + \omega_i^2)^2 \tag{7.15}$$

$\hspace{1cm}$ *where K is a positive constant and*

$$0 < \omega_1 \le \omega_2 \le \omega_3 \ldots \le \omega_l < \infty \tag{7.16}$$

(v) *For every m, $m = 1, 2, \ldots, l$, the odd polynomial n_2 in (7.13) possesses at least m purely imaginary zeros, $jb_1, jb_2, \ldots, jb_m$ such that*

$$0 < b_i \le \omega_m \qquad i = 1, 2, \ldots, m \tag{7.17}$$

Corollary 7.1 (Fujisawa) *For $Z(p)$ to be realizable as the low-pass mid-series ladder shown in Fig. 7.5(b), the necessary and sufficient conditions are those of Theorem 7.1 but in (v) n_2 is replaced by n_1.*

We now discuss the mid-shunt case of Fig. 7.5(a), and the mid-series realization of Fig. 7.5(b) will follow in a like manner. The detailed proof of sufficiency of Fujisawa's theorem is available in the references. Here, only an outline of the synthesis technique is given, which naturally constitutes an outline of the proof of sufficiency. The methodology used here is due to Youla.[28]

To begin with, we assume that the function in (7.13) is such that

$$n_1 n_2 \ne 0 \tag{7.18}$$

This is because if n_2 is identically zero, Z has all its poles on the $j\omega$-axis, and therefore the realization degenerates into a lossless one-port in series with a resistor, as explained in Section 7.4. On the other hand, if $n_1 = 0$, it is possible to show that the entire impedance degenerates into a capacitor in

126

parallel with a resistor. Therefore, to avoid these degeneracies, we assume that neither n_1 nor n_2 is identically zero and (7.18) is satisfied.

Now, the necessity of conditions (i) to (iv) is obvious, particularly (7.15) which demands that all the finite zeros of transmission must be on the $j\omega$-axis. However, the necessity of condition (v) does not become evident by a simple qualitative argument. In fact this is the crucial condition for realizability in the required structure. We now attempt to realize an impedance which satisfies conditions (i) to (v) of Theorem 7.1.

Case 1 $Z(p)$ possesses a pole at $p = \infty$. This can be extracted as a series inductor, leaving a p.r. remainder, which can be shown to satisfy the same conditions of the theorem.

Case 2 $Z(p)$ has a pole at $p = j\omega_{i0}$, $\omega_{i0} > 0$. Here $p = j\omega_{i0}$ is a zero of D in (7.13), and from (7.15) this pole must also be one of the zeros of transmission ω_i. This pole can be extracted from Z by a parallel LC in series, leaving a p.r. remainder W such that deg $W = \deg Z - 2$. Again, it is possible to show that the remainder still satisfies Theorem 7.1.

Case 3 $Z(p)$ is devoid of poles along the entire $j\omega$-axis, infinity included.

By hypothesis (condition (iii)) Z must have either a pole or a zero at $p = \infty$, and having no pole it must have a zero at $p = \infty$. Hence the admittance $Y = Z^{-1}$ has a pole at $p = \infty$.

This is the crucial point in the synthesis, to produce a ladder structure realizing the finite zeros of transmission on the $j\omega$-axis. The essence of this procedure consists in *partial* extraction of the pole at $p = \infty$ of Y, in such a way that the remainder impedance has a pole at $p = j\omega_{i0}$ where ω_{i0} coincides with one of the transmission zeros ω_i in (7.15). This pole can then be extracted as a parallel LC producing the mid-shunt combination. The cycle may then be repeated by keeping the pole at $p = \infty$ of Y in hand, and only extract it *partially* at every cycle in such a way as to *uncover* another pole of the impedance at $p = j\omega_{i0}$ where ω_{i0} is one of the finite zeros of transmission ω_i in (7.15).

Now, the partial extraction of the pole at $p = \infty$ of Y leads to the required realization in Fig. 7.5(a) *only if it obeys a certain rule*. This is now explained. Suppose that we extract a shunt capacitor C from the admittance $Y = Z^{-1}$ leaving a remainder

$$W^{-1} = Z^{-1} - Cp \tag{7.19}$$

Any value of C less than or equal to the residue of Z^{-1} at $p = \infty$, results in the remainder W^{-1} being p.r. However, C must be chosen such that W has a pole at $j\omega_i$, where ω_i is a finite zero of transmission, i.e. a zero of (7.15). The rule for this choice is as follows:

$$C = \min \left\{ \frac{Z^{-1}(j\omega_1)}{j\omega_1}, \frac{Z^{-1}(j\omega_2)}{j\omega_2}, \ldots, \frac{Z^{-1}(j\omega_l)}{j\omega_l}, \lim_{p \to \infty} \frac{Z^{-1}(p)}{p} \right\} \tag{7.20}$$

where the minimization is taken only over those quantities $Z^{-1}(j\omega_i)/j\omega_i$ which are well-defined and non-negative.

It can be shown that the rule in (7.20) for choosing C uncovers a pole of $W(p)$ at $j\omega_i$, a zero of transmission. After the extraction of this pole as a parallel LC in series, the remainder still satisfies Theorem 7.1, and the cycle may be repeated leading to the realization in Fig. 7.5(a). We also note that, in a recent contribution, the finer details of the present theorem have been considered by Fialkow.[34] His paper is very instructive and illustrates further the depth and power of Fujisawa's theorem.

Finally, the synthesis technique for a function satisfying Fujisawa's theorem: Corollary 7.1 for the mid-series ladder of Fig. 7.5(b), can be obtained in a dual manner. In every cycle in the synthesis, it is necessary to choose a series inductor such that its extraction (only partially) uncovers a pole of Y which is also a zero of transmission. The rule for choosing the partially extracted series inductor is given by

$$L = \min\left\{\frac{Z(j\omega_1)}{j\omega_1}, \frac{Z(j\omega_2)}{j\omega_2}, \ldots, \frac{Z(j\omega_l)}{j\omega_l}, \lim_{p\to\infty}\frac{Z(p)}{p}\right\} \tag{7.21}$$

It will be seen in a later chapter that Fujisawa's theorem may be applied to the synthesis of filters with optimum amplitude response.

Example 7.2 Show that the p.r. impedance

$$Z(p) = \frac{5p^3 + 5p^2 + 2p + 1}{p^4 + p^3 + 3p^2 + 2p + 1}$$

satisfies Fujisawa's theorem: Corollary 7.1 for the mid-series ladder and realize $Z(p)$.

Solution Write

$$Z(p) = \frac{(5p^2 + 1) + (5p^3 + 2p)}{(p^4 + 3p^2 + 1) + (p^3 + 2p)}$$

$$= \frac{N}{D} = \frac{m_1 + n_1}{m_2 + n_2}$$

Then

$$\tfrac{1}{2}(ND_* + N_*D) = m_1 m_2 - n_1 n_2 = (2p^2 + 1)^2$$

Now, $Z(p)$ is p.r., has a zero at $p = \infty$ and the zero of transmission is at $p_{01} = \pm j1/\sqrt{2}$. The zeros of n_1 are at $p_{02} = \pm j\sqrt{\frac{2}{5}}$. Thus $\omega_{02} < \omega_{01}$ and the impedance satisfies Fujisawa's theorem: Corollary 7.1 for the mid-series ladder.

We now proceed with the synthesis of Z. It has a zero at $p = \infty$, hence Y has a pole at this point. This is extracted as a shunt capacitor leaving a remainder Y_1 given by

$$Y_1 = Y - \tfrac{1}{5}p$$

(note $\lim_{p \to \infty} (Y/p) = \frac{1}{5}$, the residue of Y at the pole). Hence

$$Y_1 = \frac{13p^2 + 9p + 5}{25p^3 + 25p^2 + 10p + 5}$$

Next, $Z_1 = Y_1^{-1}$ has a pole at $p = \infty$; write

$$Z_1 = 5 \frac{(5p^3 + 2p) + (5p^2 + 1)}{(13p^2 + 5) + 9p}$$

At the zero of transmission $p_{01} = j1/\sqrt{2}$ we have

$$\frac{Z_1(j1/\sqrt{2})}{j1/\sqrt{2}} = 5/3$$

and

$$\lim_{p \to \infty} \frac{Z_1}{p} = \frac{25}{13}$$

Thus the pole of Z_1 at $p = \infty$ is partially extracted by a series inductor whose value is obtained according to (7.21)

$$L = \min \left\{ \frac{Z_1(j1/\sqrt{2})}{j1/\sqrt{2}}, \lim_{p \to \infty} \frac{Z_1}{p} \right\}$$
$$= \min \{5/3, 25/13\}$$
$$= 5/3 \text{ H}$$

The extraction leaves a remainder,

$$Z_2 = Z_1 - \tfrac{5}{3}p$$
$$= 5 \frac{2p^3 + 6p^2 + p + 3}{30p^2 + 27p + 15}$$
$$= 5 \frac{(2p^2 + 1)(p + 3)}{30p^2 + 27p + 15}$$

Hence $Z_2^{-1} = Y_2$ has a pole at $p = \pm j1/\sqrt{2}$, the zero of transmission, as required. This can be extracted by a series LC in shunt. Write

$$Y_2 = Z_2^{-1} = \frac{1}{5} \frac{30p^2 + 27p + 15}{(2p^2 + 1)(p + 3)}$$

The residue of Y_2 at $p = j1/\sqrt{2}$ is evaluated from (2.56) as

$$Y_2 \big|_{p \to j1/\sqrt{2}} \to \frac{2k_2 p}{(p^2 + \frac{1}{2})} \bigg|_{p \to j1/\sqrt{2}}$$

Hence $2k_2 = 9/10$ and we extract a series LC in shunt from Y_2 which leaves the remainder

$$Y_3 = Y_2 - \frac{(9/10)p}{(p^2 + \frac{1}{2})}$$
$$= \frac{3}{p + 3}$$

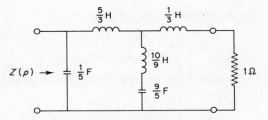

Fig. 7.6 Network of Example 7.2.

or

$$Z_3 = Y_3^{-1} = \frac{p}{3} + 1$$

and the last reactive element is a $\frac{1}{3}$ H inductor. The terminating resistor of $1\,\Omega$ is reached and the synthesis of the function is completed, as shown in Fig. 7.6.

PROBLEMS

7.1 Consider the impedance

$$Z(p) = \frac{2p^3 + 2p^2 + 2p + 1}{2p^2 + 2p + 1}$$

(a) Show that $Z(p)$ is a p.r.f.
(b) Calculate the even-part of $Z(p)$, hence determine the locations of the zeros of transmission.
(c) Realize $Z(p)$ as a resistor-terminated lossless two-port, without the use of coupled coils or ideal transformers.

7.2 A lossless two-port has the transducer power gain

$$|S_{21}(j\omega)|^2 = \frac{1}{1 + \omega^6}$$

with $1\,\Omega$ reference resistors. Calculate the input impedance of the $1\,\Omega$-terminated lossless two-port and find the realization.

7.3 Realize the transducer power gain (referred to $1\,\Omega$ resistors),

$$|S_{21}(j\omega)|^2 = \frac{1}{1 + \omega^2(4\omega^2 - 3)^2}$$

as that of a lossless ladder.

7.4 Realize the transfer function

$$S_{21}(p) = \frac{p^5}{1 + 3.236p + 5.236p^2 + 5.236p^3 + 3.236p^4 + p^5}$$

as the transmission coefficient of a $1\,\Omega$ doubly terminated lossless two-port.

7.5 Show that the driving-point impedance

$$Z(p) = \frac{5p^2 + 4p + 8}{6p^3 + 5p^2 + 16p + 8}$$

satisfies Fujisawa's Theorem for the mid-shunt ladder and find the realization.

PART III

The Augmented Theory: Synthesis of Commensurate Distributed and Digital Networks

OUTLINE

The essential features of the central theme of Part II are used to develop the synthesis techniques of two more categories of networks. In Chapter 8, passive network theory is augmented by introducing extra circuit elements of the distributed type, hence suitable for use at microwave frequencies. A coherent theory emerges which applies to the synthesis of distributed-parameter networks employing commensurate lossless transmission lines. Chapter 9 deals with digital filters with emphasis on wave digital structures which imitate the optimum sensitivity properties of passive filters. It is shown that there exists a formal analogy between digital filter transfer functions and the transfer functions of passive commensurate distributed networks.

Chapter 8

Synthesis of Commensurate Distributed Networks

8.1 INTRODUCTION

This chapter is concerned with the synthesis of some classes of distributed microwave networks. At microwave and optical frequencies, the conventional lumped elements cannot be realized, and one must seek other types of elements. These must include components capable of energy storage and transmission, but energy dissipation can still be modelled by resistors. A systematic and rigorous theory for the analysis and synthesis of these networks must then be formulated.

We have seen in Part II, how a complete theory of network synthesis may be formulated for a specific category of networks. We begin by selecting and defining a set of building blocks and regard them as admissible. Then we derive the mathematical properties of networks formed by the arbitrary interconnection of the building blocks. The external behaviour of these networks, relative to a number of ports, is described by functions and matrices. Finally, a synthesis technique is obtained for a function or matrix satisfying the appropriate realizability conditions.

Now, networks composed of various types of waveguide structures, including transmission lines, are widely used. However, if we attempt to formulate a general theory of distributed networks including all types of waveguides and transmission lines, we quickly reach an impasse. This is due to many reasons, notably the fact that such networks are describable by function and matrices of *several complex variables*; this is in contrast with lumped networks which are described by functions and matrices of *one complex variable*. Consequently the present state of the most general types of distributed networks is still in its infancy, and progress in this area has been very limited. In fact only the special cases of microwave networks which may be described by functions and matrices of one complex variable, have reached a sufficiently advanced analytical state so as to allow a fairly coherent picture to be presented.

In the present chapter we study the synthesis of two types of distributed networks.

(a) Networks composed of lossless transmission lines propagating only the transverse electromagnetic mode (TEM).[37] Moreover the lines are of equal length, or their lengths are integral multiples of the same basic length. The transmission lines are said to be *commensurate* to the same length. Such a basic length of line is called a *unit element* (UE).

(b) Coupled-line networks employing the lossless TEM multiwire line.[38] This consists of n wires together with a ground, all of the same length, each wire being coupled continuously to some or all other lines, and the surrounding dielectric is homogeneous.

Now, although the above classes of microwave networks appear to be rather restricted, many microwave components and systems can be designed on their basis. Examples are filters, broad-band transformers, broad-band delay lines, interdigital networks, and directional couplers. In fact augmenting the domain of the conventional circuit elements by the unit element *only*, leads to a much wider spectrum of sophisticated synthesis techniques. These rely very heavily on the network-theoretic ideas of lumped networks discussed in Part II. Moreover, we shall see that the defined classes of distributed networks can be treated in a unifying manner which may be regarded as a generalization and extension of the synthesis of passive lumped networks.

8.2 THE UNIT ELEMENT (UE)

Consider the section of uniform lossless transmission line of characteristic impedance Z_0 shown in Fig. 8.1. Viewed as a two-port, this basic length of line is called a *unit element* (UE). The port variables are related by

$$\begin{bmatrix} V_1 \\ I_1 \end{bmatrix} = \begin{bmatrix} \cos \beta l & jZ_0 \sin \beta l \\ jY_0 \sin \beta l & \cos \beta l \end{bmatrix} \begin{bmatrix} V_2 \\ -I_2 \end{bmatrix} \tag{8.1}$$

where $j\beta$ is the propagation function, and l is the *commensurate* length. By *commensurate* is meant that the network considered, contains line sections with the same basic length l (or integral multiples of l). Here, $\beta = \omega/c$, where c is the velocity of propagation of electromagnetic waves in the medium. We may also write

$$\tau = l/c \tag{8.2}$$

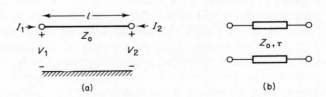

Fig. 8.1 The unit element (UE) and its symbol. (a) unit element (UE), (b) symbol of UE.

and

$$\omega\tau = \beta l \tag{8.3}$$

where τ is the commensurate one-way delay of the UE. Hence, using (8.3) in (8.1), the UE has the transmission matrix

$$[T_{UE}] = \begin{bmatrix} \cosh j\tau\omega & Z_0 \sinh j\tau\omega \\ Y_0 \sinh j\tau\omega & \cosh j\tau\omega \end{bmatrix} \tag{8.4}$$

Generalizing (8.4) to complex frequencies one obtains

$$[T_{UE}] = \begin{bmatrix} \cosh \tau p & Z_0 \sinh \tau p \\ Y_0 \sinh \tau p & \cosh \tau p \end{bmatrix} \tag{8.5}$$

Defining a new complex variable λ as

$$\lambda = \tanh \tau p$$
$$= \Sigma + j\Omega \tag{8.6}$$

the transmission matrix (8.5) of the UE may be written as

$$[T_{UE}] = \frac{1}{(1-\lambda^2)^{1/2}} \begin{bmatrix} 1 & Z_0\lambda \\ Y_0\lambda & 1 \end{bmatrix} \tag{8.7a}$$

$$= \frac{1}{(1-\lambda^2)^{1/2}} [t_{UE}] \tag{8.7b}$$

where $[t_{UE}]$ is the polynomial transmission matrix of the UE (unnormalized).

The scattering matrix of the UE, referred to equal normalizing resistors Z_0, can be obtained using (8.7) and the relationship between $[S]$ and $[t]$ in (5.53). This gives

$$[S_{UE}] = \frac{1}{(1+\lambda)} \begin{bmatrix} 0 & (1-\lambda^2)^{1/2} \\ (1-\lambda^2)^{1/2} & 0 \end{bmatrix} \tag{8.8}$$

It is sometimes convenient to use the variable

$$z = \frac{1+\lambda}{1-\lambda} = e^{2\tau p} \tag{8.9}$$

i.e.

$$\lambda = \frac{z-1}{z+1} = \frac{1-z^{-1}}{1+z^{-1}} \tag{8.10}$$

Substitution for λ from (8.10) into (8.7) and (8.8), gives

$$[T_{UE}] = \frac{1}{2z^{1/2}} \begin{bmatrix} (z+1) & Z_0(z-1) \\ Y_0(z-1) & (z+1) \end{bmatrix} \tag{8.11}$$

and

$$[S_{UE}] = \begin{bmatrix} 0 & z^{-1/2} \\ z^{-1/2} & 0 \end{bmatrix} \tag{8.12}$$

8.3 THE POSITIVE REAL CONDITION

The λ-variable defined by (8.6) is known as *Richards' variable*.[39] Its use allows the synthesis of commensurate distributed networks to be formulated on an analogous basis[40] to the synthesis of lumped networks studied in Part II in terms of the usual complex frequency variable p. We now establish the fundamental properties of driving-point impedances of commensurate distributed networks, with the UE as a starting-point.[37]

From (8.7) the input impedance of a UE when terminated in a load Z_L is

$$Z_1(\lambda) = \frac{Z_0\lambda + Z_L}{Y_0 Z_L \lambda + 1} \tag{8.13}$$

Hence, the input impedance of a UE when terminated in a short-circuit $(Z_L = 0)$ is given by

$$Z_{sc} = Z_0\lambda \tag{8.14}$$

which is called a *short-circuited stub*. When a UE is terminated in an open-circuit $(Z_L \to \infty)$, its input impedance is given by

$$Z_{oc} = 1/Y_0\lambda \tag{8.15}$$

and the result is an *open-circuited stub*.

Let us now examine the properties of a one-port formed using: short-circuited stubs, open-circuited stubs, and UEs in the cascade mode. First, the driving-point impedance $Z(\lambda)$ of a network made up of elements of impedances $Z_{0i}\lambda$ and $1/Y_{0i}\lambda$, is clearly a rational function of λ. Secondly, if a UE of characteristic impedance Z_0 is cascaded with such a network, the new input impedance $\hat{Z}$ is given by

$$\hat{Z} = Z_0 \frac{Z(\lambda) + Z_0\lambda}{Z_0 + \lambda Z(\lambda)} \tag{8.16}$$

which is also a rational function of λ. By induction, the driving-point impedance of a network containing these types of elements is a rational function of λ. These basic building blocks are shown in Fig. 8.2.

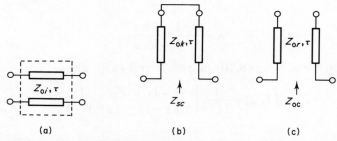

Fig. 8.2 The basic building blocks for a class of commensurate distributed networks. (a) The UE as a two-port element, (b) short-circuited stub, (c) open-circuited stub.

Now consider the properties of the λ-variable. Our objective is to show that a positive real function of λ preserves the same property with respect to the usual complex frequency variable p.

Write

$$\lambda = \Sigma + j\Omega$$
$$= \tanh\{\tau(\sigma + j\omega)\}$$
$$= \frac{\tanh \tau\sigma + j \tan \tau\omega}{1 + j \tanh \tau\sigma \tan \tau\omega} \tag{8.17}$$

from which

$$\Sigma + j\Omega = \frac{(1 + \tan^2 \tau\omega) \tanh \tau\sigma + j(1 - \tanh^2 \tau\sigma) \tan \tau\omega}{1 + \tanh^2 \tau\sigma \tan^2 \tau\omega} \tag{8.18}$$

The above expression shows that the right half λ-plane maps on to the right half p-plane. Similarly for the left halves of the two planes. Also the real and imaginary axes of the λ-plane map on to the corresponding ones in the p-plane. This mapping is shown in Fig. 8.3. It is also evident that in the case of the imaginary axes: $\sigma = 0$, $\Sigma = 0$ (representing the sinusoidal steady state)

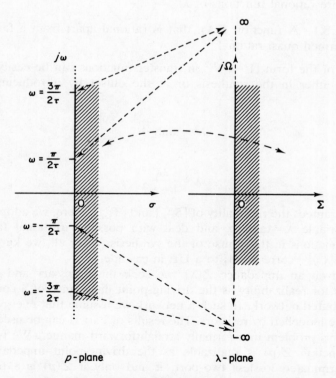

Fig. 8.3 The mapping between the p- and λ-planes.

$\lambda \to j\Omega = j \tan \tau \omega$ and the correspondence is periodic,

$$-\infty \le \Omega \le \infty \quad \text{for} \quad -\frac{(2r+1)\pi}{2} \le \omega\tau \le \frac{(2r+1)\pi}{2}$$

$$r = 0, 1, 2, \ldots \tag{8.19}$$

It follows that all responses of commensurate distributed networks are inherently periodic with respect to ω. Due to the properties of the mapping between the p-plane and λ-plane, it is clear that a positive real function of λ has the same property with respect to p.

Returning now to the basic building blocks of Fig. 8.2 we have seen that the impedance of a short-circuited stub is $Z_0\lambda$. Therefore, it has the same *formal* properties with respect to λ as those of a lumped inductor with respect to p. Similarly, the analogy between the open-circuited stub of impedance $1/Y_0\lambda$ and the capacitor is obvious. However, the UE when connected in the cascade mode, has no lumped counterpart. In fact, expression (8.8) shows that the transmission coefficient of the UE has a numerator of the form $(1-\lambda^2)^{1/2}$, corresponding to a 'half-order' transmission zero at $\lambda = \pm 1$. This results in the possible appearance of irrational factors of the form $(1-\lambda^2)^{1/2}$ in the numerators of *transfer* functions. But *driving-point* functions are rational functions of λ.

Definition 8.1 A function $F(\lambda)$, that is rational apart from a factor $(1-\lambda^2)^{1/2}$ is termed quasi-rational.

Factors of the form $(1-\lambda^2)^{1/2}$ in transfer functions can be easily accommodated, either in the synthesis or at the outset by introducing a new variable

$$\lambda' = \tanh \frac{\tau p}{2} \tag{8.20}$$

i.e.

$$\lambda = \frac{2\lambda'}{1 + \lambda'^2} \tag{8.21}$$

which guarantees the rationality of $[S_{\text{UE}}]$ and $[T_{\text{UE}}]$. Here, we adopt the use of the variable $\lambda = \tanh \tau p$, and deal with possible irrational factors in transfer functions in the course of the synthesis. After all, we know that a factor $(1-\lambda^2)^{1/2}$ corresponds to a UE in cascade.

Now, given an impedance $Z(\lambda)$, we seek the necessary and sufficient conditions for realizability as the driving-point impedance of a commensurate distributed network. In such a network we assume that energy dissipation can be modelled by resistors. The results of Part II can be used directly to solve this problem in a formally straightforward manner. We have seen that a function $Z(p)$ is realizable as the driving-point impedance of a resistor-terminated lossless two-port, if and only if $Z(p)$ is a ·p.r.f. The lossless two-port is composed entirely of elements of frequency dependence

Lp, 1/*Cp*, and ideal transformers. But we have just established the formal analogy between the short-circuited stub and the inductor, and between the open-circuited stub and the capacitor. It has also been shown that a p.r.f. of λ preserves the same property with respect to *p*. Thus we have the following result.

Theorem 8.1 *Given a rational function* $Z(\lambda)$ *which is a p.r.f. of the Richards' variable* λ, *it can always be realized as the driving-point impedance of a resistor-terminated lossless commensurate distributed two-port. The two-port is composed entirely of elements having frequency dependence* $Z_{0i}\lambda$, $1/Y_{0i}\lambda$ *and including ideal transformers (see Fig. 8.4)*

The above theorem formally establishes the sufficiency of the p.r. condition for the realizability of an arbitrary p.r.f. in the variable λ. This allows us to accommodate commensurate distributed networks into passive network theory in a manner analogous to lumped networks. We can, therefore, draw upon the wealth of results obtained in Part II by simply replacing *p* by λ in the theorems and conclusions acquired by the use of the positive real concept. However, a little reflection reveals a number of important points.

(a) Whatever happened to the UE in the cascade mode? Theorem 8.1 does not seem to make use of this possibility!

(b) Consider the general procedure for extracting imaginary-axis poles from a given $Z(\lambda)$ and its reciprocal, in a process similar to the Brune preamble (see Fig. 6.3). The structures resulting from successive extraction of stubs realizing zeros of transmission at $\lambda = 0$ and $\lambda = \infty$, lead to difficulties in manufacture. This is because a number of connections would have to be made to the same physical point, which is impossible to achieve in practice. Therefore, a means must be sought to separate the stubs without affecting the response of the network. We shall see that this can be achieved by inserting *redundant* UEs in cascade to, physically, separate the stubs without contributing to the synthesis. The same difficulties arise from attempting to realize finite $j\Omega$-axis poles by means of double stubs (equivalent to parallel or series LC's).

(c) Having succeeded in realizing all $j\Omega$-poles, including those at $\lambda = 0$ and ∞, the remaining impedance is formally realizable by the technique of cascade synthesis discussed in Chapter 6, if we replace *p* by λ and allow

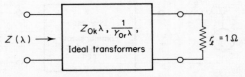

Fig. 8.4 Realization of an arbitrary rational positive real impedance.

ideal transformers. However, the ideal transformer in the microwave range is only useful as a theoretical device, not as a plausible physical element. Therefore, we should seek networks capable of realizing the microwave zero-sections (Brune, C, and D) which completely avoid the use of ideal transformers. These realizations do exist[40] and will be mentioned later.

We shall see that most of the difficulties mentioned above can be resolved using the UE in one way or another, which makes it a very powerful and versatile design tool. But first we discuss further ramifications of the use of the λ-variable in the description of networks.

8.4 THE DOUBLY TERMINATED LOSSLESS COMMENSURATE TWO-PORT

As in lumped network synthesis, we are mainly interested in the synthesis of a lossless commensurate distributed two-port which operates between a resistive source and a resistive load as shown in Fig. 8.5. The specifications on the response of the network are translated into mathematical properties of the transmission coefficient $S_{21}(\lambda)$ of the required lossless two-port. Again, the derivation of $S_{21}(\lambda)$ is the subject of approximation theory, to be discussed separately in a later chapter.

Guided by Theorem 8.1 and the results of Chapters 4 and 5, we can state the properties of the scattering matrix $[S(\lambda)]$ of a lossless commensurate distributed two-port, as shown in Fig. 8.5.

(i) Passivity and losslessness require $[S(\lambda)]$ to be unitary on the $j\Omega$-axis, i.e.

$$[S(-j\Omega)]'[S(j\Omega)]=[U_2] \tag{8.22}$$

where $[U_2]$ is the 2×2 unit-matrix.

(ii) $S_{11}(\lambda)$ and $S_{22}(\lambda)$ are rational bounded real functions.

(iii) $S_{12}(\lambda)$ and $S_{21}(\lambda)$ are rational or quasi-rational, and analytic in $\text{Re }\lambda \geq 0$, apart from a possible branch point at $\lambda = \pm1$, i.e. a factor $(1-\lambda^2)^{1/2}$.

(iv) If the two-port is reciprocal, then $[S(\lambda)]$ is symmetric, i.e. $S_{12}(\lambda) = S_{21}(\lambda)$.

(v) From (8.22) and the following properties, $[S]$ also satisfies

$$[\underline{S}(\lambda)][S(\lambda)]=[U_2] \tag{8.23}$$

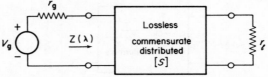

Fig. 8.5 The doubly terminated lossless commensurate distributed two-port.

where, as before the lower tilde denotes the transpose with λ replaced by $-\lambda$. Consequently expressions similar to those for the lumped case (5.29) to (5.32) are obtained by writing (8.23) explicitly. This gives

$$S_{11}S_{11*} + S_{21}S_{21*} = 1 \qquad (8.24)$$

$$S_{12}S_{11*} + S_{22}S_{21*} = 0 \qquad (8.25)$$

$$S_{11}S_{12*} + S_{21}S_{22*} = 0 \qquad (8.26)$$

$$S_{12}S_{12*} + S_{22}S_{22*} = 1 \qquad (8.27)$$

where the argument λ has been dropped for convenicnce, and the lower asterisk denotes replacing λ by $-\lambda$.

It now follows that the network-theoretic ideas underlying the synthesis of lumped networks are directly applicable to the synthesis of commensurate distributed networks. In fact we have an extra element: the UE which has no lumped counterpart; this adds to the flexibility and richness of the design techniques. Furthermore, we shall see that the UE is an indispensable element which may be used to circumvent certain practical difficulties.

The specifications on the doubly terminated two-port of Fig. 8.5 are interpreted as mathematical properties of $S_{21}(\lambda)$ or its magnitude-square $|S_{21}(j\Omega)|^2$. The synthesis procedure, then, is similar to that given in Section 6.3.2. $S_{11}(\lambda)$ is determined from (8.24) by factorization of $S_{11}S_{11*}$ and gives a bounded real function. Then the input impedance of the resistor-terminated lossless two-port is obtained as

$$Z(\lambda) = r_{\text{g}} \frac{1 + S_{11}(\lambda)}{1 - S_{11}(\lambda)} \qquad (8.28)$$

which is guaranteed positive real. We can then state that the commensurate lossless two-port is defined by the location of the zeros of transmission. These are the zeros of

$$\text{Ev } Z(\lambda) = \tfrac{1}{2}\{Z(\lambda) + Z_*(\lambda)\} \qquad (8.29)$$

Alternatively, the zeros of transmission are as defined in Theorem 5.5 with p replaced by λ.

Having gone so far in establishing the analogy and formalizing our unified approach, it is left to the reader to complete the picture and obtain the properties of $[Z(\lambda)]$, $[Y(\lambda)]$, and $[T(\lambda)]$ of a lossless commensurate distributed two-port. It is to be noted that the realization of finite zeros of transmission anywhere in the λ-plane *can be accomplished without ideal transformers*. However, this requires the use of *coupled-lines*,[40] the general properties of which will be outlined later. First we consider some special cases of practical importance which employ unit elements and stubs without requiring coupled-lines.

8.5 SYNTHESIS OF SPECIAL CONFIGURATIONS

In addition to the stubs of impedance $Z_{0i}\lambda$ and $1/Y_{0i}\lambda$, the UE in the cascade mode can be employed with great advantage from a practical as well as a theoretical viewpoint. The following theorem is of paramount importance in distributed network synthesis.

Theorem 8.2 (Richards' Theorem)[41] *Let $Z(\lambda)$ be a p.r. impedance. Then a UE of characteristic impedance $Z(1)$ can always be extracted, in cascade, from $Z(\lambda)$ leaving a remainder given by*

$$Z_1(\lambda) = Z(1)\frac{Z(\lambda) - \lambda Z(1)}{Z(1) - \lambda Z(\lambda)} \qquad (8.30)$$

that is also p.r. of degree at most equal to that of $Z(\lambda)$. Moreover, if

$$\{Ev\ Z(\lambda)\}_{\lambda=1} = 0 \qquad (8.31)$$

then

$$\deg Z_1(\lambda) = \deg Z(\lambda) - 1 \qquad (8.32)$$

The proof of Richards' theorem is given in Appendix A.4. Figure 8.6 illustrates the application of the theorem to extract a UE.

8.5.1 Cascaded unit element realizations

A consequence of Theorem 8.2 is that if *all* the zeros of Ev $Z(\lambda)$ correspond to factors of the form $(1-\lambda^2)^n$, where n is the degree of $Z(\lambda)$, then $Z(p)$ is realizable completely as a cascade of n UEs, terminated in a resistor as shown in Fig. 8.7. This is because repeated application of Richards' theorem leaves a p.r. remainder of lower degree by one than the original impedance. Hence, if deg $Z(\lambda) = n$, then after the extraction of n UEs according to (8.30), we are left with a function of zero degree, corresponding to the terminating resistor. It will be seen in a later chapter, that cascades of UEs *alone* (without stubs) can produce filtering characteristics, and are very convenient to realize in practice.

Noting that the finite zeros of transmission are those of the numerator of $S_{21}(\lambda)$ of the lossless two-port we may state the following results.

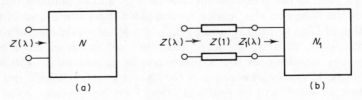

Fig. 8.6 Application of Richards' theorem. (a) One-port, (b) extraction of a UE.

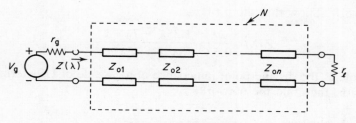

Fig. 8.7   Cascade of UEs.

Theorem 8.3   For $Z(\lambda)$ to be the driving-point impedance of a resistor-terminated cascade of n UEs as shown in Fig. 8.7, the necessary and sufficient conditions are

(i)
$$Z(\lambda) = \frac{N(\lambda)}{D(\lambda)} \text{ is a p.r.f.}$$
(8.33a)

(ii)
$$\text{Ev } Z(\lambda) = \frac{(1-\lambda^2)^n}{D_n D_{n*}}$$
(8.33b)

Corollary 8.1   For $S_{21}(\lambda)$ to be the transmission coefficient of a lossless two-port N made up of a cascade of n UEs, as shown in Fig. 8.7, the necessary and sufficient conditions are

(a)
$$|S_{21}(j\Omega)| \leq 1$$
(8.34)

(b)
$$S_{21}(\lambda) = \frac{(1-\lambda^2)^{n/2}}{P_n(\lambda)}$$
(8.35)

where $P_n(\lambda)$ is strictly Hurwitz.

Corollary 8.2   Any reactance (Foster) function $Z(\lambda)$ of degree n is realizable as the driving-point impedance of n UEs terminated in an open-circuit or a short-circuit, as shown in Fig. 8.8.

This result follows from the fact that a Foster function is a p.r.f. satisfying

$$Z(\lambda) + Z_*(\lambda) \equiv 0$$
(8.36)

Therefore, application of Richards' theorem is degree reducing. If $Z(0) = 0$ then after n extractions we reach a short-circuit, while if $Z(0) = \infty$, we reach an open-circuit.

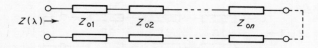

Fig. 8.8   Realization of a reactance (Foster) function.

Example 8.1   Show that the p.r.f.

$$Z(\lambda) = \frac{2\lambda^3 + 1.25\lambda^2 + 7\lambda + 1}{0.5\lambda^3 + 8\lambda^2 + 1.75\lambda + 1}$$

is realizable as the driving-point impedance of a cascade of UEs terminated in a resistor, then find the network.

Solution   Evaluating

$$\text{Ev } Z(\lambda) = \tfrac{1}{2}(Z + Z_*)$$

$$= \frac{u(\lambda)}{DD_*}$$

we obtain

$$u(\lambda) = (1 - \lambda^2)^3$$

Hence, the function satisfies (8.33). Applying Theorem 8.2,

$$Z(1) = 1 \, \Omega$$

Extracting a UE of $1 \, \Omega$ characteristic impedance we obtain the remainder given by (8.30) as

$$Z_1(\lambda) = \frac{Z(\lambda) - \lambda}{1 - \lambda Z(\lambda)}$$

which after the cancellation of a common factor $(1 - \lambda^2)$ becomes

$$Z_1(\lambda) = \frac{0.5\lambda^2 + 6\lambda + 1}{2\lambda^2 + 0.75\lambda + 1}$$

Repeating the cycle on $Z_1(\lambda)$ we evaluate $Z_1(1) = 2 \, \Omega$. Thus, a UE of $2 \, \Omega$ characteristic impedance is extracted from $Z_1(\lambda)$ leaving $Z_2(\lambda)$ given by (8.30)

$$Z_2(\lambda) = 2 \frac{Z_1(\lambda) - 2\lambda}{2 - \lambda Z_1(\lambda)}$$

$$= \frac{4\lambda + 1}{0.25\lambda + 1}$$

Finally a UE of characteristic impedance $Z_2(1) = 4 \, \Omega$ is extracted from $Z_2(\lambda)$ leaving $Z_3 = 1 \, \Omega$, which is the value of the terminating resistor. The complete realized network is shown in Fig. 8.9.

Now, repeated application of Richards' theorem, as in Example 8.1, is

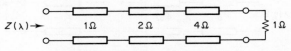

Fig. 8.9   Network of Example 8.1.

associated with problems of numerical accuracy in a high degree network. To overcome these, explicit formulae are available for the characteristic impedances of the lines in the cascade.[42] These are given below. If $Z(\lambda)$ is given then we first use the transformation in (8.10) to write

$$Z(z) = \frac{\sum\limits_{r=0}^{n} a_r z^{-r}}{\sum\limits_{r=0}^{n} b_r z^{-r}}, \quad a_0 = b_0 \tag{8.37}$$

where it is assumed that the first UE in the cascade has unity characteristic impedance. If not, a redundant dummy UE of unity characteristic impedance is cascaded in front. Next, form the upper-triangular matrices

$$[A_n] = \begin{bmatrix} a_0 & a_1 & \cdots & a_n \\ & a_0 & a_1 & \cdots & a_{n-1} \\ & & \ddots & \ddots & \vdots \\ \bigcirc & & & & a_0 \end{bmatrix} \tag{8.38}$$

$$[B_n] = \begin{bmatrix} b_0 & b_1 & \cdots & b_n \\ & b_0 & b_1 & \cdots & b_{n-1} \\ & & \ddots & \ddots & \vdots \\ \bigcirc & & & & b_0 \end{bmatrix} \tag{8.39}$$

Then, determine the matrix

$$[\phi_n] = \begin{bmatrix} 1 & c_1 & c_2 & \cdots & c_n \\ c_1 & 1 & c_1 & \cdots & c_{n-1} \\ c_2 & c_1 & 1 & \cdots & c_{n-2} \\ \vdots & & & \ddots & \\ c_n & & \cdots & & 1 \end{bmatrix} \tag{8.40}$$

from

$$2[\phi_n] = [A_n][B_n]^{-1} + [B_n][A_n]^{-1} \tag{8.41}$$

The junction reflection coefficient between the rth and $(r+1)$th UEs is given by

$$\gamma_r = (-1)^r \frac{\det \begin{bmatrix} c_1 & c_2 & \cdots & c_r \\ 1 & c_1 & \cdots & c_{r-1} \\ c_1 & 1 & \ddots & \vdots \\ c_{r-1} & \cdots & 1 & c_1 \end{bmatrix}}{\det \begin{bmatrix} 1 & c_1 & c_2 & \cdots & c_{r-1} \\ c_1 & 1 & c_1 & \cdots & c_{r-2} \\ \vdots & & & \ddots & \vdots \\ c_{r-1} & & \cdots & & 1 \end{bmatrix}} \tag{8.42}$$

Finally, the characteristic impedances of the UEs are obtained from

$$Z_{0(r+1)} = Z_{0r} \frac{1+\gamma_r}{1-\gamma_r} \qquad r = 1, 2, \ldots, n \tag{8.43}$$

The above expressions obviate the need for synthesis by repeated application of Richards' theorem and can be easily programmed on a digital computer. As pointed out before, filters can be designed using cascades of UEs alone; hence the importance of the above results.

8.5.2 All-stub networks

Given a p.r.f. $Z(\lambda)$ we consider the conditions under which it is realizable as the driving-point impedance of a resistor-terminated low-pass ladder as shown in Fig. 8.10. The series arms are short-circuited stubs and the shunt ones are open-circuited stubs. These conditions can be easily obtained by direct use of the results of Section 7.2.1 with p replaced by λ. Therefore $Z(\lambda)$ is realizable as shown in Fig. 8.10 if and only if

(a) $$Z(\lambda) = \frac{N(\lambda)}{D(\lambda)} \text{ is a p.r.f.} \tag{8.44a}$$

(b) $$\text{Ev } Z(\lambda) = \frac{K}{DD_*} \tag{8.44b}$$

where K is a constant. In other words, all the zeros of transmission are at $\lambda = \infty$. Thus, the transmission coefficient of the lossless ladder is a *bounded real* function of the form

$$S_{21}(\lambda) = \frac{1}{P(\lambda)} \tag{8.45}$$

i.e. $P(\lambda)$ is strictly Hurwitz.

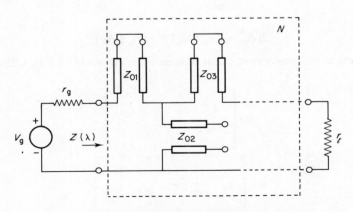

Fig. 8.10 Low-pass all-stub ladder.

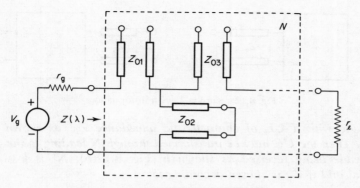

Fig. 8.11   Quasi-high-pass all-stub ladder.

The synthesis of the impedance is achieved via its continued fraction expansion as given by (7.4) with p replaced by λ.

Similarly, using the results of Section 7.2.2 with p replaced by λ a given impedance $Z(\lambda)$ is realizable as the driving-point impedance of a resistor-terminated quasi-high-pass ladder as shown in Fig. 8.11 if and only if

(a)
$$Z(\lambda) = \frac{N(\lambda)}{D(\lambda)} \text{ is a p.r.f.} \qquad (8.46a)$$

(b)
$$\text{Ev } Z(\lambda) = \frac{K\lambda^{2n}}{DD_*} \qquad (8.46b)$$

where K is a constant and n is the degree of $Z(\lambda)$. Thus, in this case, all the zeros of transmission are at $\lambda = 0$. In terms of the transmission coefficient, this must be a *bounded real* function of the form

$$S_{21}(\lambda) = \frac{\lambda^n}{P_n(\lambda)} \qquad (8.47)$$

i.e. $P_n(\lambda)$ is strictly Hurwitz.

Now, in the lumped domain there is no objection to these structures. However, for the distributed networks of Figs 8.10 and 8.11, a number of stubs must exist at a fixed point in space; but the appropriate connections cannot be achieved in practice. It is therefore desirable to introduce UEs between the stubs to effect physical separation. We shall now consider the case where the UEs do not contribute to the overall response, i.e. they are redundant in an electrical sense.

Theorem 8.4 *Consider a commensurate lossless reciprocal two-port N defined by the transmission matrix*

$$[T(\lambda)] = \begin{bmatrix} A(\lambda) & B(\lambda) \\ C(\lambda) & D(\lambda) \end{bmatrix} \qquad (8.48)$$

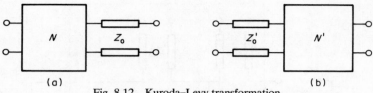

Fig. 8.12   Kuroda–Levy transformation.

in cascade with a UE of characteristic impedance Z_0, as shown in Fig. 8.12(a). Then the UE may be transformed through N leading to the order of the cascade being reversed as shown in Fig. 8.12(b). N′ is lossless and reciprocal and if Z_0' is chosen according to

$$Z_0' = \frac{A + BY_0}{C + DY_0}\bigg|_{\lambda=1} \tag{8.49}$$

then the networks of Fig. 8.12(a) and (b) are equivalent.

The proof of the above theorem is straightforward and may be found in

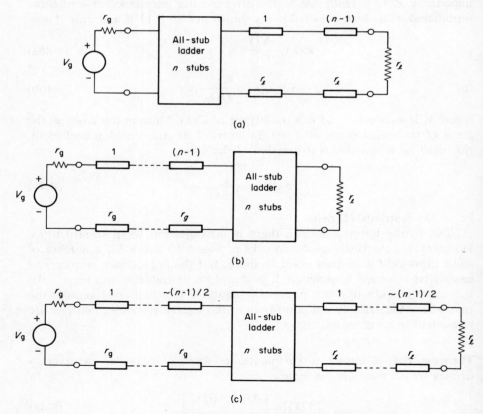

Fig. 8.13   Alternative ways of inserting redundant UEs.

the references.[43] Here we consider its application to introduce UEs to isolate the stubs in Figs 8.10 and 8.11. Insert a cascade of $(n-1)$ UEs all of the same characteristic impedance r_ℓ, between the load r_ℓ and the all-stub ladder as shown in Fig. 8.13(a). Note that introducing these UEs does not affect the input impedance, since together with the load the UEs have an input impedance equal to r_ℓ. Applying the transformation of Fig. 8.12 repeatedly, a UE can be moved through the network into any position between the stubs to yield a practical realization, in which all the stubs are separated by UEs. Another way of achieving the same objective is to insert $(n-1)$ UEs all of the characteristic impedance r_g at the source end as shown in Fig. 8.13(b). Alternatively $(n-1)/2$ UEs (or nearest integer) all of characteristic impedance r_ℓ are inserted between the load and the ladder, while the other $(n-1)/2$ are all of characteristic impedance r_g and inserted at the source end. This is shown in Fig. 8.13(c).

Having inserted the redundant UEs in any of the three possible ways of Fig. 8.13, the transformation of Fig. 8.12 can be used in the simple case where N is a single stub. The transmission matrices of the four types of stubs are given by:

(i) Series short-circuited stub

$$[T(\lambda)] = \begin{bmatrix} 1 & Z_0\lambda \\ 0 & 1 \end{bmatrix} \tag{8.49}$$

(ii) Shunt open-circuited stub

$$[T(\lambda)] = \begin{bmatrix} 1 & 0 \\ Y_0\lambda & 1 \end{bmatrix} \tag{8.50}$$

(iii) Series open-circuited stub

$$[T(\lambda)] = \begin{bmatrix} 1 & \dfrac{1}{Y_0\lambda} \\ 0 & 1 \end{bmatrix} \tag{8.51}$$

(iv) Shunt short-circuited stub

$$[T(\lambda)] = \begin{bmatrix} 1 & 0 \\ \dfrac{1}{Z_0\lambda} & 1 \end{bmatrix} \tag{8.52}$$

In order to apply the transformation of Fig. 8.12, we first write the transmission matrix of the stub–UE cascade before and after the reversal. Comparison of these matrices and use of (8.49) leads to the Kuroda transformations[44] illustrated in Fig. 8.14. It is to be noted that any ideal transformers produced by the transformation can be referred back to the load.

Example 8.2 The network shown in Fig. 8.15(a) is a microwave low-pass

150

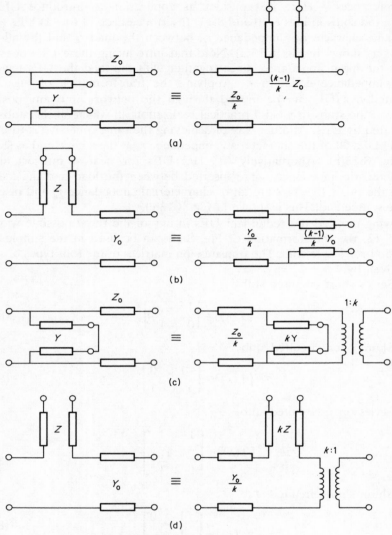

Fig. 8.14   Kuroda's transformations, $k = 1 + YZ_0$ or $k = 1 + Y_0 Z$.

filter, derived from a $1\,\Omega$-terminated lumped prototype by letting $p \to \lambda$. Use the Kuroda transformations in Fig. 8.14 to introduce UEs between the stubs.

Solution   There are three possible ways to proceed; we consider two and the third is left to the reader. (i) Insert two cascaded UEs of $1\,\Omega$ characteristic impedance between the load and the all-stub ladder, as shown in Fig. 8.15(b). Use Fig. 8.14 to move one of the UEs through the first stub. Repeat the process twice until all stubs are separated by UEs. The network of Fig. 8.15(c) is reached, which contains UEs and one type of stubs only, namely: series short-circuited. (ii) Insert a UE of 1 Ω characteristic impedance at the load end, and another at the source end as shown in Fig. 8.15(d). Apply Kuroda's transformations of Fig. 8.14 twice to obtain the network of Fig. 8.15(e). This contains two UEs and one type of stubs only, namely: shunt open-circuited. This is the most desirable structure from the practical viewpoint.

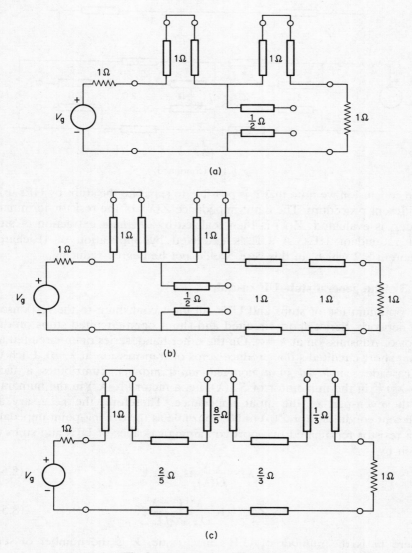

Fig. 8.15   Networks of Example 8.2.

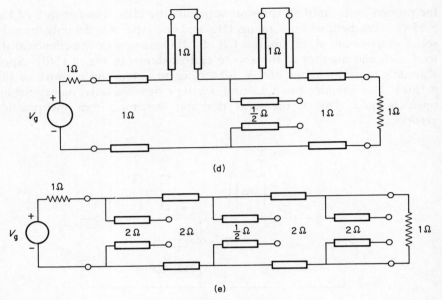

Fig. 8.15 (continued)

In conclusion we note that it is possible to separate the stubs by UEs using a different procedure. The input impedance $Z(\lambda)$, of the resistor-terminated ladder, is evaluated. $Z(\lambda)$ is then realized by *alternate* extraction of stubs and redundant UEs. A UE is extracted by application of (Richards') Theorem 8.2 which, in this case, would *not* be degree-reducing.

8.5.3 The general stub–UE cascade

For optimum use of stubs and UEs, all must contribute to the response of the network. Series short-circuited and shunt open-circuited stubs produce zeros of transmission at $\lambda = \infty$. On the other hand, series open-circuited and shunt short-circuited stubs produce zeros of transmission at $\lambda = 0$. Each UE in cascade, *employed in a non-redundant manner*, contributes a factor $(1-\lambda^2)^{1/2}$ in the numerator of $S_{21}(\lambda)$, i.e. a factor $(1-\lambda^2)$ in the numerator of the even-part of the input impedance. Therefore, the necessary and sufficient conditions for $Z(p)$ to be realizable as the driving-point impedance of a resistor-terminated two-port containing cascades of UEs and stubs are given by

(i) $$Z(\lambda) = \frac{N(\lambda)}{D(\lambda)} \text{ is a p.r.f.} \tag{8.53a}$$

(ii) $$\text{Ev } Z(\lambda) = K \frac{\lambda^{2k}(1-\lambda^2)^m}{D_{m+k+l}D_{m+k+l*}} \tag{8.53b}$$

where m is the number of U.E.s in cascade, k is the number of series open-circuited and shunt short-circuited stubs, l is the number of series short-circuited and shunt open-circuited stubs, and K is a constant.

In terms of the transmission coefficient $S_{21}(\lambda)$, it is necessary and sufficient that

(i)
$$|S_{21}(j\Omega)| \le 1 \tag{8.54a}$$

(ii)
$$S_{21}(\lambda) = \frac{\lambda^k(1-\lambda^2)^{m/2}}{P_{m+k+l}(\lambda)} \tag{8.54b}$$

where $P(\lambda)$ is strictly Hurwitz. The synthesis of the impedance is accomplished by extracting poles at $\lambda = 0$ and $\lambda = \infty$ from Z or Y corresponding to the stubs, while the U.E.s are extracted by application of Richards' theorem according to (8.30).

8.6 COUPLED-LINE NETWORKS

8.6.1 Description and general properties[38,45]

We now give a brief discussion of lossless commensurate distributed networks employing the coupled-line or TEM multiwire line section. This is shown in Fig. 8.16, and consists of n wires (together with a ground) all of the same length, each wire being coupled continuously to every other line throughout its length, and the surrounding dielectric is homogeneous. There are n-ports at each end (between each wire and ground) and the resulting $2n$-port becomes a circuit element. For the n-wire line shown in Fig. 8.16, let

C_{ie} = capacitance per unit length to ground of ith wire (8.55a)
$C_{ik} = C_{ki}$ = capacitance per unit length between the ith and
  kth wires $(i \ne k)$ (8.55b)
L_{ii} = self-inducatance per unit length of the ith wire (8.55c)
$L_{ik} = L_{ki}$ = mutual inductance between the ith and kth wires
  $(i \ne k)$ (8.55d)

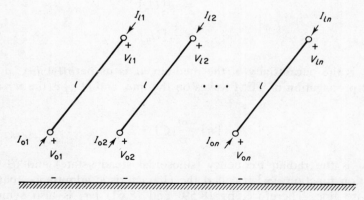

Fig. 8.16 The n-wire line.

Define

$$C_{ii} = \sum_{\substack{k=1 \\ k \neq i}}^{n} C_{ik} + C_{ie} \qquad (8.56)$$

and the matrices

$$[C] = \begin{bmatrix} C_{11} & -C_{12} & \cdots & -C_{1i} & \cdots & -C_{1n} \\ -C_{12} & C_{22} & & & & \vdots \\ \vdots & & & & & \\ -C_{1i} & & & & & \\ \vdots & & & & & \\ -C_{1n} & & \cdots & & & C_{nn} \end{bmatrix} \qquad (8.57)$$

$$[L] = \begin{bmatrix} L_{11} & L_{12} & \cdots & L_{1i} & \cdots & L_{1n} \\ L_{12} & L_{22} & & & & \vdots \\ \vdots & & & & & \\ L_{1i} & & & & & \\ \vdots & & & & & \\ L_{1n} & & \cdots & & & L_{nn} \end{bmatrix} \qquad (8.58)$$

In (8.57) $[C]$ is the static capacitance matrix of the n-wire line. It is symmetric with positive diagonal entries and negative off-diagonal entries. From (8.55) to (8.57), $[C]$ is symmetric hyper-dominant, i.e.

$$C_{ki} = C_{ik} \geq 0 \qquad i \neq k \qquad (8.59a)$$

$$C_{ii} - \sum_{\substack{k=1 \\ k \neq i}}^{n} C_{ik} = C_{ie} \geq 0 \qquad (8.59b)$$

In (8.58) $[L]$ is the static inductance matrix of the n-wire line, and is symmetric with positive elements.

Now for the defined TEM n-wire line, $[L]$ and $[C]$ satisfy

$$[L][C] = [C][L]$$
$$= \varepsilon\mu[U_n]$$
$$= \frac{1}{c^2}[U_n] \qquad (8.60)$$

where ε is the permittivity of the medium, μ is its permeability, c is the speed of propagation of TEM waves on the line, and $[U_n]$ is the $n \times n$ unit matrix.

Let

$$[\eta] = \frac{\omega}{\beta}[C] \qquad (8.61)$$

where ω is the radian frequency (sinusoidal steady state) and $j\beta$ is the propagation function. $[\eta]$ is called the characteristic admittance matrix of the n-wire line. Therefore, by (8.59) and (8.61) $[\eta]$ is also symmetric hyper-dominant, with $\eta_{ik} = \dfrac{\omega}{\beta} C_{ik}$,

$$\eta_{ki} = \eta_{ik} \geq 0 \qquad i \neq k \tag{8.62a}$$

$$\eta_{ii} - \sum_{\substack{k=1 \\ k \neq i}}^{n} \eta_{ik} = \eta_{ie} \geq 0 \tag{8.62b}$$

We can also define the characteristic impedance matrix of the n-wire line as

$$[\xi] = \frac{1}{(\varepsilon\mu)^{1/2}} [L] \tag{8.63}$$

Combining (8.60), (8.61), and (8.63) we obtain

$$[\eta][\xi] = [U_n] \tag{8.64}$$

Viewed as a $2n$-port, the port voltages and currents of the n-wire line are

$$[V_0] = \begin{bmatrix} V_{01} \\ V_{02} \\ \vdots \\ V_{0n} \end{bmatrix}, \quad [V_l] = \begin{bmatrix} V_{l1} \\ V_{l2} \\ \vdots \\ V_{ln} \end{bmatrix} \tag{8.65a}$$

$$[I_0] = \begin{bmatrix} I_{01} \\ I_{02} \\ \vdots \\ I_{0n} \end{bmatrix}, \quad [I_l] = \begin{bmatrix} I_{l1} \\ I_{l2} \\ \vdots \\ I_{ln} \end{bmatrix} \tag{8.65b}$$

Then, these matrices can be related in a number of ways. With the definitions of $[\eta]$ and $[\xi]$ we may write

$$\begin{bmatrix} [V_0] \\ [I_0] \end{bmatrix} = \frac{1}{(1-\lambda^2)^{1/2}} \begin{bmatrix} [U_n] & [\xi]\lambda \\ [\eta]\lambda & [U_n] \end{bmatrix} \begin{bmatrix} [V_l] \\ -[I_l] \end{bmatrix} \tag{8.66}$$

where λ is Richards' variable. The above expression describes the external behaviour of the $2n$-port formed from the n-wire line. Alternatively we may obtain the following descriptions of the $2n$-port,

$$\begin{bmatrix} [I_0] \\ [I_l] \end{bmatrix} = \frac{1}{\lambda} \begin{bmatrix} [\eta] & -[\eta](1-\lambda^2)^{1/2} \\ -[\eta](1-\lambda^2)^{1/2} & [\eta] \end{bmatrix} \begin{bmatrix} [V_0] \\ [V_l] \end{bmatrix} \tag{8.67}$$

and

$$\begin{bmatrix} [V_0] \\ [V_l] \end{bmatrix} = \frac{1}{\lambda} \begin{bmatrix} [\xi] & [\xi](1-\lambda^2)^{1/2} \\ [\xi](1-\lambda^2)^{1/2} & [\xi] \end{bmatrix} \begin{bmatrix} [I_0] \\ [I_l] \end{bmatrix} \tag{8.68}$$

It is to be noted that conditions (8.62) have been shown to be necessary and sufficient for the existence of the general n-wire line.

Various special cases of the n-wire line may be constructed to form networks capable of performing a wide variety of tasks. For a detailed discussion of these and for a comprehensive list of references, the reader may consult a recent review paper.[38] Here, we only point out some special cases of practical importance.

8.6.2 Special cases of coupled-line networks

First consider the basic interdigital line[46-48] shown in Fig. 8.17. This consists of an n-wire line where coupling exists *only between adjacent lines*, and alternate wires are short-circuited to ground at opposite ends. The other ends of the wires are open-circuited, except for the first and last wires which provide the input and output ports. The characteristic admittance matrix of the line is given by

$$[\eta] = \begin{bmatrix} (\eta_1 + \eta_{12}) & -\eta_{12} & 0 & \ldots \ldots 0 \\ -\eta_{12} & (\eta_2 + \eta_{12} + \eta_{23}) & -\eta_{23}0 & \ldots \ldots 0 \\ 0 & -\eta_{23} & (\eta_3 + \eta_{23} + \eta_{34}) & \ldots 0 \\ \vdots & & & \vdots \\ & & & -\eta_{n-1,n} \\ 0 & \ldots & 0 & -\eta_{n-1,n} & (\eta_n + \eta_{n-1,n}) \end{bmatrix}$$

$$(8.69)$$

It is then possible to show that the resulting two-port has the equivalent network of Fig. 8.18 which contains only uncoupled lines. This is a cascade of UEs and stubs. Moreover, the transformations in Fig. 8.14 can be used to combine all the stubs into a single shunt short-circuited stub. This reveals that the two-port possesses only a single zero of transmission at $\lambda = 0$, while the remaining zeros are produced by the $(n-1)$ UEs. Therefore, the transmission coefficient of the basic interdigital network must be of the form

$$S_{21}(\lambda) = \frac{\lambda(1-\lambda^2)^{(n-1)/2}}{P_n(\lambda)} \qquad (8.70)$$

where $P_n(\lambda)$ is strictly Hurwitz.

A generalized interdigital network[49] can be obtained if we allow coupling to exist between all lines in Fig. 8.17. The advantage of these lines is that they afford the possibility of realizing more general classes of transfer functions, than the restricted one of (8.70).

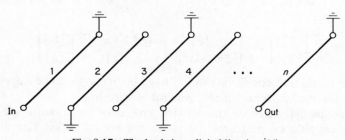

Fig. 8.17   The basic interdigital line (n odd).

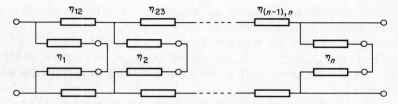

Fig. 8.18   Equivalent network of the basic interdigital line.

Finally, a widely-used microwave component is the multi-element directional coupler shown in Fig. 8.19. The network is designed such that: with equal terminating resistors on all four ports and using port 1 as input, port $2l$ is perfectly isolated whereas the other ports are perfectly matched.[38]

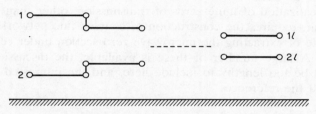

Fig. 8.19   The multi-element directional coupler.

8.7   GENERAL CONSIDERATIONS—CASCADE SYNTHESIS

Let the starting-point in the synthesis be $S_{21}(\lambda)$ or $|S_{21}(j\Omega)|^2$ of the lossless commensurate two-port, required to operate between a resistive source and a resistive load, as shown in Fig. 8.5. Then, the first step is to form the input impedance $Z(p)$ of the resistor-terminated lossless two-port as explained in Section 8.4. A realizable $S_{21}(\lambda)$ will lead to a p.r. $Z(\lambda)$, and we have demonstrated, in a formal way, the sufficiency of the p.r. condition. However, the need for ideal transformers presents a serious obstacle which must be overcome. To this end, a complete synthesis technique has been presented for any p.r. impedance $Z(\lambda)$ as the driving-point impedance of a resistor-terminated lossless commensurate *transformerless* two-port employing *coupled-lines*.[40]

We now give a recapitulation of the main points in the present chapter and point out various extensions for the interested reader. Consider the following possibilities which arise according to the locations of the zeros of transmission.

(i) Ev $Z(\lambda) = $ a constant. Then, $Z(\lambda)$ is realizable as a lossless one-port in series with a resistor. This follows directly from the result of Section 7.4 if we replace p by λ. Furthermore, by Corollary 8.1, the lossless

158

one-port itself is realizable as a cascade of UEs terminated in a short-circuit or an open-circuit.

(ii) Zeros of transmission resulting from factors of the form $(1-\lambda^2)^m$ in Ev $Z(\lambda)$ or $(1-\lambda^2)^{m/2}$ in $S_{21}(\lambda)$, can be extracted as cascades of UEs at any convenient point during the synthesis. This is a consequence of Theorem 8.2.

(iii) Zeros of transmission at $\lambda = 0$ and $\lambda = \infty$ are realized by appropriate stubs, whereas degenerate $j\Omega$-axis zeros of transmission (i.e. those which are also poles of the entire impedance Z or the admittance Y) can be realized by double stubs. These are extracted in a process similar to the Brune preamble in the lumped case (see Section 6.4 and Fig. 6.3). The reactance functions, themselves, removed during the Brune preamble can also be realized as open-circuited or short-circuited cascade of UEs.

(iv) The realization of finite zeros of transmission, other than those in (ii) and (iii) requires the construction of coupled-line networks which are capable of extracting the microwave zero-section under certain conditions. A large number of these is available, the discussion of which would be too lengthy to include here, and the interested reader may consult the references.[40,45,46]

8.8 NON-RECIPROCAL REALIZATIONS—REFLECTION NETWORKS

For a reciprocal realization of a transmission coefficient $S_{21}(\lambda)$, its numerator must be even or odd, apart from a possible factor $(1-\lambda^2)^{1/2}$. If not, augmentation is necessary, and the realized network has an increased degree. Moreover, after augmentation coupled-line structures are generally required for the realization. Sometimes non-reciprocal realizations are sought, which may result in simple structures.[8] In certain cases this objective can be achieved using the ideal three-port circulator shown in Fig. 8.20. Its scattering matrix (see Problem 5.5 and Fig. P5.5) is given by

$$[S] = \begin{bmatrix} 0 & 0 & e^{j\psi_1} \\ e^{j\psi_2} & 0 & 0 \\ 0 & e^{j\psi_3} & 0 \end{bmatrix} \tag{8.71}$$

Fig. 8.20 Three-port ideal circulator.

The circulator is non-reciprocal and can be built using microwave devices employing ferrites. Examination of (8.71) reveals that all the diagonal entries are zero, implying zero reflection coefficients at all ports. Thus all ports are matched. Relating the incident and reflected signals $[\alpha]$ and $[\beta]$ at the ports by $[S]$ we have

$$\begin{bmatrix} \beta_1 \\ \beta_2 \\ \beta_3 \end{bmatrix} = [S] \begin{bmatrix} \alpha_1 \\ \alpha_2 \\ \alpha_3 \end{bmatrix} \tag{8.72}$$

Therefore, using (8.71) in (8.72) we have,

$$\beta_1 = \alpha_3 e^{j\psi_1}, \qquad \beta_2 = \alpha_1 e^{j\psi_2}, \qquad \beta_3 = \alpha_2 e^{j\psi_3} \tag{8.73}$$

This means that when a signal is incident at port i ($i = 1, 2, 3$) with all other ports matched, none is reflected, none is transmitted to port $(i-1)$, but all of it is transmitted without loss to port $(i+1)$, apart from a phase shift. Hence, the three-port has a cyclic power transmission property, and

$$|S_{13}| = |S_{21}| = |S_{32}| = 1 \tag{8.74}$$

In the simplest case all the angles in (8.71) are zero and the non-zero scattering parameters are unity. However, more generally, each of the non-zero transmission coefficients is an all-pass function.

Now if a one-port N_2 with reflection coefficient $S_{11}(\lambda)$ is connected at port 2 of the circulator with matching resistors at ports 1 and 3, a two-port N results operating between ports 1 and 3 as shown in Fig. 8.21. The scattering matrix of the two-port N is easily seen to be

$$[\hat{S}] = \begin{bmatrix} 0 & 1 \\ S_{11} & 0 \end{bmatrix} \tag{8.75}$$

except for a constant phase shift. Therefore the *transmission* coefficient of N is the same as the *reflection* coefficient of N_2. The input impedance of the one-port N_2 is given by

$$Z_2(\lambda) = \frac{1 + \hat{S}_{21}}{1 - \hat{S}_{21}} \tag{8.76}$$

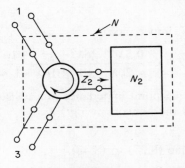

Fig. 8.21 Reflection network.

and if

$$\hat{S}_{21}(\lambda) = \frac{N_m(\lambda)}{D_n(\lambda)} \tag{8.77}$$

then

$$\text{Ev } Z_2(\lambda) = \frac{DD_* - NN_*}{(D-N)(D_* - N_*)} \tag{8.78}$$

When the numerator of Ev Z_2 is either a constant or of the form $K\lambda^{2n}$, then Z_2 is realizable as a resistor-terminated all-stub ladder. Similarly if $\hat{S}_{21}(\lambda)$ is an all-pass function, Ev $Z_2 = 0$ and Z_2 is a reactance function realizable as a cascade of UEs terminated in an open-circuit or a short-circuit. Thus, simple non-reciprocal realizations can be obtained in these special cases. The overall network N is called a *reflection network*.[8]

8.9 CONCLUSION

Our approach to the synthesis of commensurate distributed networks has relied on the use of Richards' variable $\lambda = \tanh \tau p$. This allowed the theory to be formulated by following, very closely, the analytical lines previously developed in Part II for lumped networks in terms of the usual complex frequency variable p. It has been shown that a given p.r.f. $Z(\lambda)$ is realizable as the driving-point impedance of a resistor-terminated lossless commensurate two-port. We have pointed out that, even in the most general case, the lossless two-port may be realized without transformers.

The study has concentrated on the properties and use of the unit element (UE), and a brief discussion of coupled-line networks was also given. At this point, a coherent theory of passive network synthesis has emerged, which encompasses both lumped networks and very useful classes of distributed networks. This enlargement of the arsenal of passive circuit elements leads to a remarkable enrichment of the design techniques.

PROBLEMS

8.1 Show that the p.r. impedance

$$Z(\lambda) = \frac{0.5\lambda^3 + 26\lambda^2 + 8.5\lambda + 10}{20\lambda^3 + 8\lambda^2 + 16\lambda + 1}$$

is realizable as the driving-point impedance of a resistor-terminated cascade of UEs and hence find the network using:

(a) Richards' theorem.
(b) The explicit formulae (8.37) to (8.43).

8.2 Realize the following reactance function, once as an all-stub ladder and a second time as a cascade of UEs terminated in a short-circuit or an open-circuit.

$$Z(\lambda) = \frac{\lambda^4 + 6\lambda^2 + 8}{\lambda^5 + 8\lambda^3 + 15\lambda}$$

8.3 Realize the following p.r. impedance

$$Z(\lambda) = \frac{\lambda^3 + 3.5\lambda^2 + 6\lambda}{\lambda^3 + 4\lambda^2 + 6\lambda + 10}$$

8.4 (a) Form the input impedance of the $1\,\Omega$-terminated lossless two-port whose transducer power gain is given by

$$|S_{21}(j\Omega)|^2 = \frac{1}{1 + 10\,\Omega^6}$$

(b) Realize the driving-point impedance obtained in (a) as a $1\,\Omega$-terminated all-stub ladder.

(c) Use the Kuroda transformations to introduce UEs to achieve physical separation of the stubs in the realization obtained in (b).

8.5 Consider the transfer function

$$S_{21}(\lambda) = \frac{\lambda(1 - \lambda^2)}{\lambda^3 + 2\lambda^2 + 2\lambda + 1}$$

Show that $S_{21}(\lambda)$ is realizable as the transmission coefficient of an interdigital network. Realize the function using the equivalent circuit of the interdigital line (Fig. 8.18).

Chapter 9
Synthesis of Digital Filters

9.1 INTRODUCTION

Digital signal processing has become an important technique in a large number of applications which were reviewed in Chapter 1. This field can be divided into two main areas, the first is digital filtering and the second is spectrum analysis. Attention in this chapter, and throughout the book, is devoted entirely to the first area namely, the synthesis of digital filters. Therefore the words *filter* and *network* are used synonymously. We derive the general properties of the transfer functions used to characterize digital filters and discuss the realization techniques of these transfer functions, employing the basic digital hardware building blocks. The problem of determining the transfer function such that it satisfies prescribed specifications is the subject of a later chapter on approximation theory.

Digital filters can be used directly when the signals to be processed are already in digital form. Alternatively, in the case of analogue signals, the digital filter must be preceded by an analogue-to-digital (A/D) converter. Digital-to-analogue (D/A) can also be performed at the output of the digital filter if the signals are required in analogue form. This general situation is depicted in Fig. 9.1 where the entire network, including the converters is said to be of the *sampled-data* type. Naturally A/D conversion requires sampling, quantization, and encoding, while D/A conversion involves quantization and decoding. The reader may consult the references for a discussion of A/D and D/A conversion.[50]

In line with the general philosophy of this book, our approach to the realization techniques of digital filters, is hardware-oriented. However, digital filters are often implemented using microprocessors and computers, and the actual realizations are chiefly software implementation. The reader who studies the present chapter, should find no difficulty in adapting the techniques to any particular processor-oriented design. The study of computers and microprocessors, is nowadays part of all electrical engineering curricula.

9.2 THE z-TRANSFORM

In the description of sampled-data networks, we deal with discrete functions of time by contrast with continuous ones. Such functions are defined only at

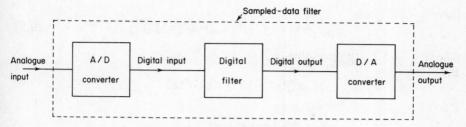

Fig. 9.1 The digital filter in analogue environment.

discrete instants which are integral multiples of the same quantity: T seconds. Consider a function defined as

$$f(t) \triangleq f(rT) \qquad r = 0, 1, 2, \ldots \tag{9.1}$$

Such a function may be thought of as a sequence of numbers $\{f(0), f(T), f(2T), \ldots\}$ representing the values of the function at the instants 0, T, $2T, \ldots$. In the notation of (9.1), $f(rT)$ denotes the *entire* sequence, not just the value at the rth instant. More accurately, one should write $\{f(rT)\}$ with the curled brackets denoting a sequence, but these are usually dropped for convenience. When dealing with number sequences, a useful mathematical tool is the z-transform, a detailed discussion of which is available in many references.[16,51] Here, we only give the definition of the standard one-sided z-transform, together with some of its most important properties. It is assumed henceforth that $f(t)$ in (9.1) is zero for $t < 0$, and the sequence $f(rT)$ is therefore zero for $r < 0$. In this case the one-sided and the double-sided z-transforms of the sequence are the same.

Consider a sequence of numbers $f(rT)$; its standard one-sided z-transform $F(z)$ is defined as

$$F(z) = \mathscr{Z}\{f(rT)\}$$
$$= \sum_{r=0}^{\infty} f(rT)z^{-r} \tag{9.2}$$

where

$$z = e^{Tp} \tag{9.3a}$$

or

$$z^{-1} = e^{-Tp} \tag{9.3b}$$

which is the *unit delay operator*. The following are some important properties of the z-transform.

P.1   The z-transform is linear, i.e. if

$$\mathscr{Z}\{f_1(rT) = F_1(z) \tag{9.4a}$$

and

$$\mathscr{Z}\{f_2(rT)\} = F_2(z) \tag{9.4b}$$

then

$$\mathscr{Z}\{af_1(rT) + bf_2(rT)\} = aF_1(z) + bF_2(z) \tag{9.5}$$

where a and b are constants.

P.2

$$\mathscr{Z}\{f(rT - kT)\} = z^{-k}F(z) \tag{9.6}$$

P.3

$$\mathscr{Z} \sum_{k=0}^{r} f_1(kT)f_2(rT - kT)$$

$$= \mathscr{Z} \sum_{k=0}^{r} f_1(rT - kT)f_2(kT)$$

$$= F_1(z) \cdot F_2(z) \tag{9.7}$$

A useful function is the discrete-time unit impulse defined as

$$\delta(rT) = 1 \qquad r = 0$$
$$= 0 \qquad r \neq 0 \tag{9.8}$$

and its z-transform, from (9.2), is unity.

9.3 THE SAMPLED-DATA FILTER

Consider the diagram of Fig. 9.2(a) which shows the process of periodic sampling of a continuous signal. After (ideal) sampling, the signal is defined at discrete instants and this operation can be viewed simply as sequence generation, as shown in Fig. 9.2(b). Accordingly, if a signal $x(t)$ is sampled every T seconds, the sampled version can be represented as a train of impulses given by

$$x(rT) = \sum_{k=0}^{\infty} x(kT)\delta(rT - kT) \tag{9.9}$$

where $\delta(rT - kT)$ is a unit impulse at $t = kT$. Again, $x(rT)$ denotes the entire sequence $\{x(0), x(T), x(2T), \ldots\}$ not just the rth sample. It is also implied that $x(t)$ is zero for $t < 0$. The sampled signal may then be quantized and encoded, and is usually thought of in terms of a sequence of numbers $x(rT)$ representing the values of the signal at the sampling instants. Clearly the sampling (Nyquist) frequency is

$$f_N = \frac{1}{T} = \frac{\omega_N}{2\pi} \tag{9.10}$$

which, by the sampling theorem, *just* exceeds twice the highest frequency of the bands of interest. In other words, in order for the continuous signal $x(t)$ to be recoverable unambiguously from its samples $x(rT)$, the signal $X(j\omega)$ must be band-limited to $< \omega_N/2$, ω_N being the radian sampling frequency taken here as the *minimum* required value.

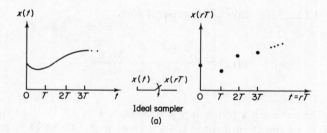

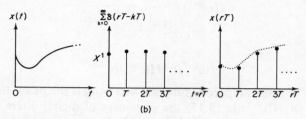

Fig. 9.2 The sampling process. (a) Sampling a continuous
signal, (b) sequence generation.

Now let the input sequence to the digital filter be $x(rT)$ and the output
sequence be $y(rT)$. It is then possible to relate the input and output
sequences by a difference equation, with constant coefficients, of the general
form

$$y(rT) = \sum_{i=0}^{m} a_i x(rT - iT) - \sum_{i=1}^{n} b_i y(rT - iT) \qquad (9.11)$$

The above expression is simply a recursion formula, allowing the present
value of the output to be calculated from: the n past output values, the m
past input values, and the present input value.

Taking the z-transform of both sides of (9.11) and making use of (9.2) to
(9.7) we obtain

$$Y(z) = X(z) \sum_{i=0}^{m} a_i z^{-i} - Y(z) \sum_{i=1}^{n} b_i z^{-i} \qquad (9.12)$$

Thus, a transfer function $H(z)$ of the digital filter may be formed as

$$H(z) = \frac{Y(z)}{X(z)} \qquad (9.13)$$

and using (9.12), $H(z)$ takes the general form

$$H(z) = \frac{\sum_{i=0}^{m} a_i z^{-i}}{1 + \sum_{i=1}^{n} b_i z^{-i}} = \frac{M(z^{-1})}{N(z^{-1})} \tag{9.14}$$

Since z and the complex frequency p are related by (9.3), then the transfer function $H(z)$ of the digital filter is such that: if A/D and D/A conversions are performed at the input and output respectively, then

$$H(e^{j\omega T}) = \frac{\sum_{i=0}^{m} a_i e^{-ji\omega T}}{1 + \sum_{i=1}^{n} b_i e^{-ji\omega T}} \tag{9.15}$$

or

$$H(e^{j\omega T}) = |H(e^{j\omega T})| \, e^{j\psi(\omega)} \tag{9.16}$$

where $|H(e^{j\omega T})|$ is the amplitude function and $\psi(\omega)$ is the phase function. Due to the form of $H(e^{j\omega T})$ in (9.15) the responses of digital filters are periodic with respect to ω, the period being $\omega_N = 2\pi f_N$.

9.4 RICHARDS' VARIABLE IN THE DIGITAL DOMAIN

Although the variable z appears naturally in the transfer functions of digital filters, it is often more convenient to use a different variable in the description of digital filters. This is particularly true in the solution of the approximation problem, i.e. the derivation of the transfer function such that it possesses filtering characteristics meeting the required specifications. This variable is defined as

$$\lambda = \tanh \frac{T}{2} p \tag{9.17}$$

which is a bilinear function of z, i.e. from (9.3)

$$\lambda = \frac{1 - z^{-1}}{1 + z^{-1}} \tag{9.18}$$

or

$$z^{-1} = \frac{1 - \lambda}{1 + \lambda} \tag{9.19}$$

Clearly λ is the Richards' variable which was introduced in the previous chapter to describe commensurate distributed networks, with a slightly different interpretation of the argument, i.e. $T/2$ replaces τ in (8.6). Therefore, we may still call λ in (9.17) Richards' variable, where

$$\lambda = \tanh \tau p \tag{9.20}$$

and τ is the one-way delay of the unit element in the distributed domain, but in the digital domain τ is half the sampling period!

Using (9.19) and (9.14), a digital filter transfer function assumes the alternative form

$$H(\lambda) = \frac{P(\lambda)}{Q(\lambda)} \tag{9.21}$$

where $P(\lambda)$ and $Q(\lambda)$ are polynomials in Richards' variable. *This immediately establishes the formal analogy between digital and commensurate distributed transfer functions.* It follows that digital transfer functions become immediately available once we know (or obtain) solutions to the approximation problem in the λ-domain. This allows us to use the wealth of material available in the area of commensurate distributed network synthesis. The full significance of this analogy will be apparent when we consider approximation theory in a later part of the book.

Let us now examine the mapping between the p, λ, and z^{-1} planes. Write (9.17) as

$$\lambda = \Sigma + j\Omega$$

$$= \tanh \frac{T}{2}(\sigma + j\omega) \tag{9.22a}$$

Therefore

$$\Sigma + j\Omega = \frac{\left(1 + \tan^2 \frac{T}{2}\omega\right)\tanh\frac{T}{2}\sigma + j\left(1 - \tanh^2\frac{T}{2}\sigma\right)\tan\frac{T}{2}\omega}{1 + \tan^2\frac{T}{2}\omega \tanh^2\frac{T}{2}\sigma} \tag{9.22b}$$

As in the case of commensurate distributed networks, (9.22) shows that points in the right half p-plane map on to points in the right half λ-plane. Similarly the left half-planes map on to each other. The real and imaginary axes map on to the corresponding ones. In the case of the imaginary axes (of particular interest since they represent the sinusoidal steady state) $\sigma = 0$, $\Sigma = 0$ and the correspondence is periodic. Thus, for $p = j\omega$, $\lambda = j\Omega = j\tan(T/2)\omega$, and (9.10) gives

$$\frac{T}{2}\omega = \left(\frac{\omega}{\omega_N}\right)\pi = \left(\frac{f}{f_N}\right)\pi \tag{9.23}$$

with ω_N being the radian sampling frequency, so that

$$-\infty \leq \Omega \leq \infty \quad \text{for} \quad \frac{-(2r+1)}{2} \leq \frac{\omega}{\omega_N} \leq \frac{(2r+1)}{2} \tag{9.24}$$

$$r = 0, 1, 2, \ldots$$

But the sampling theorem limits the useful frequency range by

$$0 \leq \omega < \frac{\omega_N}{2} \tag{9.25}$$

so that a continuous signal is recoverable unambiguously from its samples if it is band-limited to the range $<\omega_N/2$, outside which aliasing occurs. It is, therefore, assumed henceforth that the maximum usable frequency is half the sampling frequency though, in fact, it is just below $\omega_N/2$.

Returning to (9.22) it is an easy matter to show that the imaginary axis of the p-plane (hence that of the λ-plane) maps on to the unit circle of the z^{-1}-plane. The right half p-plane (hence the right half λ-plane) maps on to the interior of the unit circle in the z^{-1}-plane. Similarly the left half p-plane (hence the left half λ-plane) maps on to the exterior of the unit circle in the z^{-1}-plane. The mapping between the three planes is shown in Fig. 9.3.

Now it must be remembered that one is always interested in the properties of functions in terms of the true complex frequency variable p. In particular, our study concentrates on network responses in the sinusoidal steady state $(p = j\omega)$. Stability of the network requires that its transfer function given by (9.14) or (9.21) must have all its poles in the open left half p-plane. This corresponds to the open left half λ-plane and the exterior of the unit circle in the z^{-1}-plane. Therefore (as established also in Chapter 8) a Hurwitz polynomial in Richards' variable λ preserves the same property with respect to p. Hence the use of the variable λ allows the use of the simple test for the Hurwitz character of $Q(\lambda)$ in (9.21) and the stability of the transfer function is checked in a straightforward manner.

In the light of the above discussion, the clear-cut simple correspondence between the p-plane and the λ-plane makes Richards' variable the most convenient tool for studying the properties of digital filters, and for finding solutions to the approximation problem, i.e. derivation of a realizable transfer function which meets certain specifications. Moreover, the form of a digital transfer function $H(\lambda)$ is identical to an $S_{21}(\lambda)$ of a commensurate distributed filter with the same response if we observe that $T/2$ in the digital case replaces τ in the distributed case. So the solutions to the approximation problem differ only with regard to the non-reciprocal nature of the digital

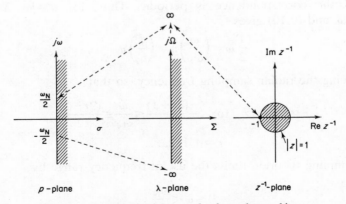

Fig. 9.3 The mapping between the three planes of interest.

filter. The essential difference is, however, in the realization. Naturally if the derivation of the transfer function is accomplished in the λ-domain, while the realization requires the function to be expressed in terms of z^{-1}, one can substitute for λ from (9.18) into (9.21) to re-express the transfer function in terms of the required variable.

9.5 STABILITY AND CLASSIFICATION OF DIGITAL FILTERS

For stability, the digital filter transfer function as expressed in (9.14) must be devoid of poles in $|z^{-1}| \le 1$, corresponding to the closed right half p-plane. This, in turn, corresponds to the closed right half λ-plane, and $Q(\lambda)$ in (9.21) must be a strictly Hurwitz polynomial in Richards' variable λ. This is the only realizability condition which a digital transfer function must satisfy, and can be easily checked using the test in Section 2.5, expansion (2.50), with p replaced by λ. However, a digital filter is inherently *non-reciprocal* and can readily be active.

In a broad sense, the methods of realization of a digital transfer function are of two types: recursive and non-recursive. For a recursive realization the relationship between the input sequence $x(rT)$ and the resulting output sequence $y(rT)$ is given by

$$y(rT) = F\{y(rT - T), y(rT - 2T), \ldots, x(rT), x(rT - T), x(rT - 2T), \ldots\}$$
(9.26)

which means that the present output sample $y(rT)$ is a function of past outputs as well as present and past input samples. For a non-recursive realization,

$$y(rT) = F\{x(rT), x(rT - T), x(rT - 2T), \ldots\}$$
(9.27)

i.e. the present output sample is a function only of past and present inputs.

Next consider a different way of classifying digital filters. If in (9.14) $N(z^{-1}) = 1$, i.e.

$$H(z) = \sum_{r=0}^{m} a_r z^{-r}$$
(9.28)

then the filter has a *finite duration impulse response* (FIR) and is said to be of the FIR-type. It is inherently stable since $H(z)$ is devoid of poles in the z^{-1}-plane. Such a filter is also capable of providing *exact* linear phase response at *all* frequencies, as will be shown in a later chapter. The realization of an FIR transfer function is very simple and can be done either recursively or non-recursively. The non-recursive realization of FIR filters is, however, the most commonly used type.

By contrast, if in (9.14) $N(z^{-1}) \ne 1$, i.e. it is a non-trivial polynomial the filter is of the *infinite duration impulse response* type (IIR). Again it can be realized either recursively or non-recursively although the former method is the one which is usually employed.

170

9.6 SIMPLE REALIZATION TECHNIQUES

The basic building blocks employed in the synthesis of digital filters are the adder, multiplier, and unit delay shown symbolically in Fig. 9.4. The detailed construction of these building blocks can be found in many references dealing with the design of logic circuits.[50] Naturally, these elements may be regarded as *operations* to be implemented in *software* form, using a computer or a microprocessor.

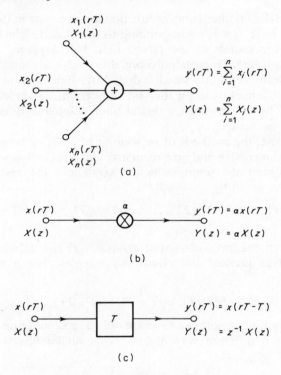

Fig. 9.4   The basic digital building blocks. (a) Adder,
(b) multiplier, (c) unit delay.

9.6.1   IIR filters

A general transfer function as expressed in (9.14) corresponds to a difference equation of the form given in (9.12). Such an equation can be implemented using multipliers, adders, and unit delays, in the direct form shown in Fig. 9.5. In this realization of equation (9.14) we have taken $m = n$ for convenience. If $m \neq n$ the appropriate paths corresponding to the zero coefficients are deleted from Fig. 9.5. Clearly this type of realization is extremely simple to obtain, since only knowledge of the coefficients of the transfer function is necessary. The network shown in Fig. 9.5 is canonic

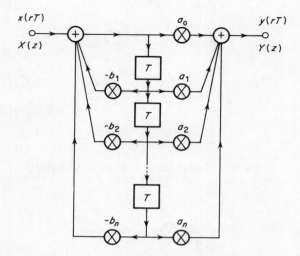

Fig. 9.5 Direct canonic recursive realization of an IIR
transfer function.

since it contains the minimum possible number of unit delays, this being
equal to the degree of the transfer function.

Despite the ease by which the realization of Fig. 9.5 may be obtained,
there is a severe coefficient sensitivity problem associated with this structure.
This implies both the deviation of the actual frequency response from the
desired one, and also a certain noise due to round-off accumulation. This
problem becomes more pronounced as the degree of the network increases.
This situation can be improved by first factoring the transfer function into
quadratic terms (as well as a possible first order term for odd-degree
functions). Each term is then realized separately and the individual networks
are connected in cascade to realize the entire transfer function. Thus the
transfer function in (9.14) is expressed as

$$H(z) = \prod_k H_k(z) \tag{9.29}$$

where a typical quadratic factor is of the form

$$H_k = \frac{\alpha_{0k} + \alpha_{1k}z^{-1} + \alpha_{2k}z^{-2}}{1 + \beta_{1k}z^{-1} + \beta_{2k}z^{-2}} \tag{9.30}$$

which can be realized as shown in Fig. 9.6. For a first order factor of (9.29)
we have

$$H_k = \frac{\alpha_{0k} + \alpha_{1k}z^{-1}}{1 + \beta_{1k}z^{-1}} \tag{9.31}$$

which can be realized as shown in Fig. 9.7. The entire network realized by
cascading the factors of the transfer function, takes the form of Fig. 9.8.

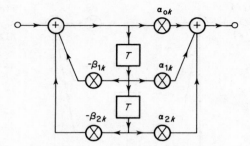

Fig. 9.6   Realization of a second-order factor
of the form (9.30).

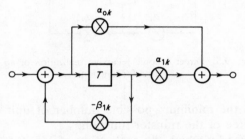

Fig. 9.7   Realization of a first-order factor of the
form (9.31).

An alternative realization can proceed by first expanding the transfer
function in partial fraction form as

$$H(z) = K + \sum_i \frac{\alpha_{0i} + \alpha_{1i}z^{-1}}{1 + \beta_{1i}z^{-1} + \beta_{2i}z^{-2}}$$

$$= K + \sum_i \hat{H}_i(z) \qquad (9.32)$$

where K is a constant. Each term in (9.32) is then realized by a network of
the same form in Fig. 9.6 which realizes (9.30) but with the path corresponding to α_{2k} deleted. Also a first order section may result if in (9.32) a
term has $\alpha_{1i} = 0$, $\beta_{2i} = 0$. This is realized as shown in Fig. 9.7 with the path
containing α_{1k} deleted (together with the output adder). The resulting
subnetworks are finally connected in the so-called parallel form shown in
Fig. 9.9.

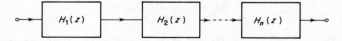

Fig. 9.8   Cascade form of realization.

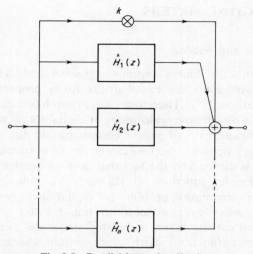

Fig. 9.9   Parallel form of realization.

9.6.2   FIR filters

A further improvement from the sensitivity viewpoint results with non-recursive realizations of FIR functions. Such a transfer function is given by (9.28), i.e. it reduces to a polynomial in z^{-1}. A simple direct realization of an FIR function is shown in Fig. 9.10, and is often called a transversal (tapped-delay line) filter.

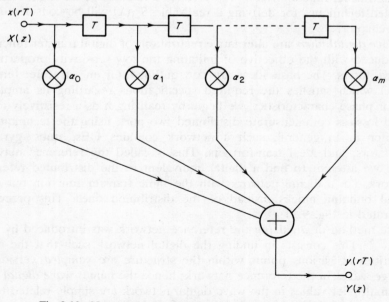

Fig. 9.10   Non-recursive realization of an FIR transfer function.

9.7 WAVE DIGITAL FILTERS

9.7.1 Motivation and outline

It is well known that the doubly terminated passive lossless filter (lumped or commensurate distributed), has excellent sensitivity properties with respect to element variations.[52–54] Therefore, attention has been given to the problem of finding alternative realizations of digital filters which imitate the low sensitivity characteristics of passive commensurate distributed networks. The formal analogy between the two categories of networks was explained in Section 9.4. It is dictated by the fact that both categories of networks are describable by transfer functions of Richards' variable λ. Consequently, solutions to the approximation problem for digital filters can be obtained in terms of λ, and many existing solutions (i.e. transfer functions) in the distributed domain can be taken to be digital transfer functions. Thus, if $S_{21}(\lambda)$ is a transfer function which gives certain characteristics, in the commensurate distributed domain, then if we reinterpret λ as $\tanh (T/2)p$ then we immediately obtain a digital transfer function as

$$H(\lambda) \triangleq S_{21}(\lambda) \qquad (9.33)$$

The magnitude and phase responses of the realized digital network will be the same as those of the corresponding commensurate distributed one, regardless of the particular method of realization. Naturally, if required, the digital transfer function can be expressed as a function of z using (9.18). The detailed techniques for deriving a realizable $S_{21}(\lambda)$ will be considered in a later chapter.

Wave digital filters are alternative realizations of digital transfer functions, introduced with the objective of imitating the low sensitivity properties of passive filters. The basic ideas are now outlined. Given a transfer function $S_{21}(\lambda)$ which satisfies the required specifications regarding its amplitude and/or phase characteristics, we begin by realizing it as a resistively terminated lossless commensurate distributed two-port, using the techniques of Chapter 8. In general, such a network contains UEs, stubs, gyrators, circulators, and ideal transformers. This is called the reference network. Then we attempt to find a digital equivalent to the distributed reference network, i.e. a digital network with the same transfer function but using digital building blocks instead of the distributed ones. This process is illustrated in Fig. 9.11.

The method of *digitizing* the reference network was introduced by Fettweis.[55,56] This consists in finding the digital network such that the signal quantities at various points within the structure are *sampled* versions of *voltage waves* in the reference network; hence the name: *wave digital filter*. The multiplier values in the wave digital network are simply related to the element values of the reference filter. In many practical cases, the excellent

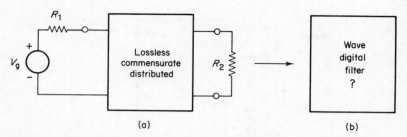

Fig. 9.11 Pertinent to the derivation of a digital filter imitating the low sensitivity properties of a passive filter. (a) Reference network, (b) wave digital equivalent.

sensitivity properties of the passive reference filter are transferred to the digital filter to produce low coefficient sensitivity.

9.7.2 Wave digital building blocks

Digitization of the reference filter is accomplished by first digitizing each element, then the individual digitized elements are interconnected preserving the same topology to obtain the desired wave digital filter. Naturally, we still employ the adder, multiplier, and unit delay, but more composite building blocks result for digitization of the passive elements. To obtain the wave digital equivalent of a typical element in the reference commensurate distributed filter, use is made of the so-called *wave flow diagrams*. These show the incident and reflected voltage waves at the various ports of a general n-port as depicted in Fig. 9.12(a). This is akin to the scattering description of networks, except that the incident and reflected quantities are *voltage waves* in the present case.

Consider the one-port shown in Fig. 9.12(b), where the steady state incident and reflected voltage waves are denoted by A and B, respectively. These are defined by

$$A = V + RI$$
$$B = V - RI$$

(9.34)

where R is a positive constant called the *port reference resistance*. If the instantaneous quantities are $v(t)$ and $i(t)$ then the instantaneous incicent and reflected waves are given by

$$a(t) = v(t) + Ri(t)$$
$$b(t) = v(t) - Ri(t)$$, $$t = rT$$

(9.35)

Henceforth, *we shall use capital letters A_i, B_i to denote steady state quantities and small letters a_i, b_i to denote instantaneous quantities.* If the input impedance of the one-port is Z, then

$$V = ZI$$

(9.36)

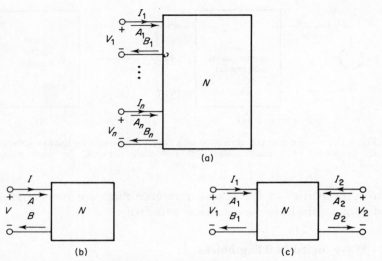

Fig. 9.12 Illustrating incident and reflected voltage waves. (a) n-port, (b) one-port, (c) two-port.

which upon substitution in (9.34) gives

$$B = \frac{Z-R}{Z+R} A$$

$$= SA \tag{9.37}$$

and as in the scattering description, S is the reflection coefficient of the one-port

In the case of the two-port shown in Fig. 9.12(c) we have for port reference resistances of R_1 and R_2,

$$A_1 = V_1 + R_1 I_1$$
$$B_1 = V_1 - R_1 I_1 \tag{9.38}$$

and

$$A_2 = V_2 + R_2 I_2$$
$$B_2 = V_2 - R_2 I_2 \tag{9.39}$$

It is understood that the instantaneous values $a_k(t)$, $b_k(t)$ are defined by (9.38), (9.39) if we replace V_k, I_k by $v_k(t)$, $i_k(t)$, where $k = 1, 2$.

Generalization of (9.36), (9.37) to the description of the n-port in Fig. 9.12(a) is straightforward and gives,

$$A_k = V_k + R_k I_k$$
$$B_k = V_k - R_k I_k \tag{9.40}$$
$$k = 1, 2, 3, \ldots, n$$

where R_k is the kth-port reference resistance.

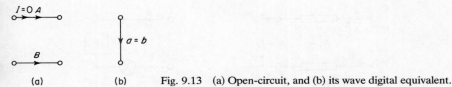

(a)   (b)   Fig. 9.13   (a) Open-circuit, and (b) its wave digital equivalent.

We now use the above type of description to obtain wave digital equivalents for the elements and conditions in a reference distributed network.

OPEN-CIRCUIT AND SHORT-CIRCUIT CONDITIONS

For an open-circuit (9.37) gives

$$B = A, \quad b = a \tag{9.41}$$

resulting in the wave flow diagram of Fig. 9.13. Similarly (9.37) gives for a short-circuit

$$B = -A, \quad b = -a \tag{9.42}$$

which results in the wave flow diagram of Fig. 9.14.

SHORT-CIRCUITED STUB

Let a short-circuited stub have a characteristic impedance R, i.e. the same as the port reference resistance. Its input impedance is $Z = R\lambda$ as shown in Section 8.2. Using (8.14) we can write

$$Z_{sc} = R\lambda = R\frac{1 - z^{-1}}{1 + z^{-1}} \tag{9.43}$$

which upon substitution in (9.37) gives

$$B = -z^{-1}A \tag{9.44a}$$

and taking the inverse z-transform we have

$$b(t) = -a(t - T), \quad t = rT \tag{9.44b}$$

The wave flow diagram of (9.44) is shown in Fig. 9.15, in which the input and output signals are clearly sampled versions of the incident and reflected waves of the short-circuited stub.

Fig. 9.14   (a) Short-circuit, and (b) its wave digital equivalent

178

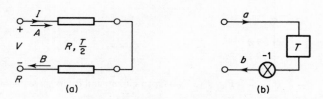

Fig. 9.15 (a) Short-circuited stub, and (b) its wave digital equivalent.

OPEN-CIRCUITED STUB

For an open-circuited stub of characteristic impedance equal to the port reference resistance R, we have

$$Z_{oc} = \frac{R}{\lambda} = R\frac{1+z^{-1}}{1-z^{-1}} \tag{9.45}$$

which upon substitution in (9.37) gives

$$B = z^{-1}A \tag{9.46a}$$

or

$$b(t) = a(t-T), \quad t = rT \tag{9.46b}$$

The reference stub and its wave digital equivalent are shown in Fig. 9.16.

UNIT ELEMENT (UE)

As given by (8.7) a UE of characteristic impedance R and delay $T/2$ as shown in Fig. 9.17(a) is described by

$$\begin{bmatrix} V_1 \\ I_1 \end{bmatrix} = \frac{1}{(1-\lambda^2)^{1/2}} \begin{bmatrix} 1 & R\lambda \\ \lambda/R & I \end{bmatrix} \begin{bmatrix} V_2 \\ -I_2 \end{bmatrix} \tag{9.47}$$

which upon use in (9.38) to (9.39) gives

$$\begin{bmatrix} B_1 \\ B_2 \end{bmatrix} = \begin{bmatrix} 0 & z^{-1/2} \\ z^{-1/2} & 0 \end{bmatrix} \begin{bmatrix} A_1 \\ A_2 \end{bmatrix} \tag{9.48}$$

Since the factor $z^{-1/2}$ corresponds to a delay of $T/2$, the wave digital equivalent of the UE is directly obtained from (9.48) as shown in Fig. 9.17(b).

Fig. 9.16 (a) Open-circuited stub, and (b) its wave digital equivalent.

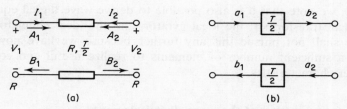

Fig. 9.17   (a) UE and, (b) its wave digital equivalent.

RESISTOR

For a resistor whose value equals the port reference resistance R, we have from (9.36) and (9.37)

$$S = 0, \quad b = 0 \tag{9.49}$$

The resistor, together with its wave flow diagram are shown in Fig. 9.18.

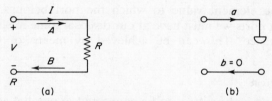

Fig. 9.18   (a) Resistor, and (b) its wave digital equivalent.

RESISTIVE SOURCE

For a voltage source V_g with real internal impedance R as shown in Fig. 9.19(a) we have

$$V_g = V + RI \tag{9.50}$$

and use of (9.50) in (9.34) results in

$$A = V_g \tag{9.51a}$$

or

$$a(t) = v_g(t), \quad t = rT \tag{9.51b}$$

The corresponding wave flow diagram is shown in Fig. 9.19(b), (c)

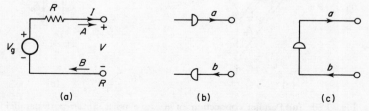

Fig. 9.19   (a) Resistive source and (b), (c) wave digital equivalents.

Finally, we note that it is also possible to derive wave digital equivalents of the ideal transformer, the ideal gyrator, and the ideal circulator. However, we shall not pursue this any further because we have now at our disposal a sufficient number of elements to realize useful and commonly used filter characteristics.

9.7.3 Interconnection of the wave digital elements

Having given the wave digital equivalents of the elements in a typical reference filter, attention is now devoted to the problem of interconnecting these building blocks to form the complete digital filter preserving the topology of the reference distributed filter. In principle, the complete network can be obtained by interconnecting the ports and sources. However, a difficulty arises in the resulting wave flow diagrams due to the assumption made about the values of the port reference resistors. In Section 9.7.2, these were not chosen arbitrarily, but rather their values were imposed by the element value to which the port belongs. Hence, when interconnecting ports we must have at our disposal some means for changing the port resistance. This can be achieved by means of three types of *adaptors*.

PARALLEL ADAPTOR

Consider a number of ports: n to be connected in parallel as shown in Fig. 9.20(a). The port variables are constrained by the parallel connection to satisfy

$$V_1 = V_2 = \ldots = V_n \tag{9.52a}$$

$$I_1 + I_2 + \ldots + I_n = 0 \tag{9.52b}$$

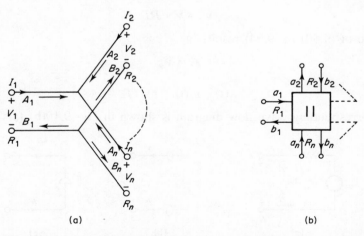

(a)   (b)

Fig. 9.20   (a) Parallel connection of n wave ports, (b) n-port parallel adaptor.

Using the above constraints in (9.40) we obtain

$$B_k = A_0 - A_k \tag{9.53}$$

where

$$A_0 = \sum_{k=1}^{n} \alpha_k A_k \tag{9.54}$$

$$\alpha_k = \frac{2G_k}{\sum\limits_{k=1}^{n} G_k}, \quad G_k = \frac{1}{R_k} \tag{9.55}$$

and R_k $(k = 1, 2, \dots, n)$ are the port reference resistors. Therefore the B_k are calculated from A_k using (9.53) to (9.55). This requires the usual collection of digital multipliers and adders. The result is an n-port device capable of performing the parallel connection of a number of ports: n. This is called an *n-port parallel adaptor* and is given the schematic symbol of Fig. 9.20(b). Clearly, if the calculations are to be made directly as in (9.53) to (9.55), the n-port parallel adaptor requires n multipliers. However, we note that (9.55) gives

$$\sum_{k=1}^{n} \alpha_k = 2 \tag{9.56}$$

and we can express A_0 in (9.54) as

$$A_0 = 2A_n + \sum_{k=1}^{n-1} \alpha_k (A_k - A_n) \tag{9.57}$$

Therefore, the calculations now require only $(n-1)$ multipliers since α_n has been eliminated. It is always possible to eliminate one multiplier, and the associated port is called the *dependent port*.

Example 9.1 Sketch the wave flow diagram of a three-port parallel adaptor with port 3 as the dependent port.

Solution In this case we use (9.57) with $n = 3$, thus

$$A_0 = 2A_3 + \sum_{k=1}^{2} \alpha_k (A_k - A_3)$$
$$= 2A_3 + \alpha_1 (A_1 - A_3) + \alpha_2 (A_2 - A_3)$$

and using (9.53) we have

$$B_1 = A_0 - A_1, \quad B_2 = A_0 - A_2, \quad B_3 = A_0 - A_3$$

which lead to the wave flow diagram shown in Fig. 9.21 together with the symbol for the three-port parallel adaptor.

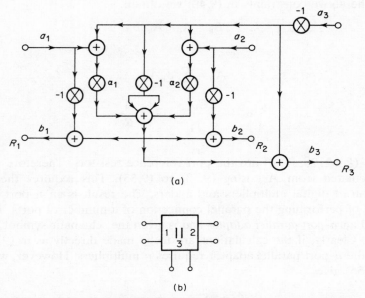

(a)

(b)

Fig. 9.21   (a) Wave flow diagram of a three-port parallel adaptor with port
3 as the dependent port, (b) symbol of a three-port parallel adaptor.

SERIES ADAPTOR

When a number of ports: n are connected in series as shown in Fig. 9.22(a)
the port variables are constrained by

$$V_1 + V_2 + \ldots + V_n = 0 \qquad (9.58)$$

$$I_1 = I_2 = \ldots = I_n \qquad (9.59)$$

Combining the above constraints with (9.40) we obtain

$$B_k = A_k - \beta_k A_0 \qquad (9.60)$$

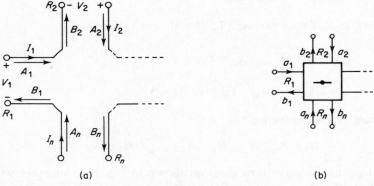

(a)   (b)

Fig. 9.22   (a) Series connection of n wave ports, (b) n-port series adaptor.

where

$$A_0 = \sum_{k=1}^{n} A_k \qquad (9.61)$$

$$\beta_k = \frac{2R_k}{\sum_{k=1}^{n} R_k} \qquad (9.62)$$

Again the calculation of B_k from A_k is performed using multipliers and adders. The resulting digital network is called an *n-port series adaptor*, and is given the schematic symbol of Fig. 9.22(b). Here, we can also reduce the number of multipliers by one if we note that (9.62) gives

$$\sum_{k=1}^{n} \beta_k = 2 \qquad (9.63)$$

Therefore, one of $\beta_k : \beta_n$, say, can be eliminated leading to

$$B_n = -A_0 - \sum_{k=1}^{n-1} B_k \qquad (9.64)$$

and B_k ($k = 1, 2, \ldots, n-1$) can be calculated using $(n-1)$ multipliers while B_n can be calculated from (9.64) which does not need a multiplier. Thus an *n*-port series adaptor requires $(n-1)$ multipliers, and the port whose α_k has been eliminated is called the *dependent* port.

Example 9.2 Sketch the wave flow diagram of a three-port series adapter with port 3 as the dependent port.

Solution Using (9.60) to (9.64), with $n = 3$ we have

$$B_1 = A_1 - \beta_1 A_0, \quad B_2 = A_2 - \beta_2 A_0,$$
$$B_3 = -A_0 - (B_1 + B_2)$$

which lead to the wave digital realization of the three-port series adapter shown in Fig. 9.23 together with its symbol.

THE CASCADE ADAPTOR: SIMPLE CHANGE OF REFERENCE RESISTANCE

For simple connection of ports with different resistances, as shown in Fig. 9.24(a) either a two-port series or parallel adapter may be used. An alternative arrangement is also possible. Consider the two ports 1 and 2 with port resistances R_1 and R_2 respectively. The waves are related by (9.38) to (9.39), i.e.

$$\left. \begin{array}{l} A_k = V_k + R_k I_k \\ B_k = V_k - R_k I_k \end{array} \right\} \quad k = 1, 2 \qquad (9.65)$$

If these ports are simply interconnected as in Fig. 9.24(a) we have

$$V_1 = V_2, \quad I_1 = -I_2 \qquad (9.66)$$

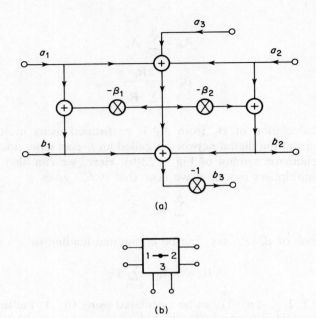

(a)

(b)

Fig. 9.23 (a) Wave flow diagram of a three-port series adaptor with port 3 as the dependent port, (b) symbol of a three-port series adaptor.

Elimination of V_k and I_k in (9.65), using (9.66), gives

$$B_1 = A_2 + \alpha \cdot (A_2 - A_1) \tag{9.67a}$$
$$B_2 = A_1 + \alpha \cdot (A_2 - A_1) \tag{9.67b}$$

where

$$\alpha = \frac{R_1 - R_2}{R_1 + R_2} \tag{9.68}$$

But the above expressions define a *wave two-port* which will be called a *cascade adaptor*. It is represented schematically by the symbol of Fig.

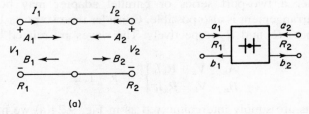

(a)

Fig. 9.24 (a) Cascade connection of two wave ports, (b) cascade adaptor.

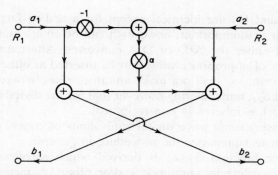

Fig. 9.25 Wave flow diagram of a cascade adaptor.

9.24(b). From (9.67) to (9.68) a detailed wave flow diagram can be drawn as shown in Fig. 9.25 giving a possible realization which requires one multiplier.

9.7.4 Wave digital equivalents of some commensurate distributed structures

Consider Fig. 9.26 which shows a reference passive filter together with its wave digital equivalent. Let the input to the wave digital filter be taken at the wave digital equivalent of the source. Also let the output be taken at the port to which the wave digital equivalent of the resistive load is connected. If the forward transmission coefficient of the reference filter is $S_{21}(\lambda)$ then

$$S_{21}(\lambda) = 2\sqrt{\frac{R_1}{R_2}}\frac{V_2}{V_g} \qquad (9.69)$$

But the *voltage wave transfer function* of the digital filter is obtained from (9.38) and (9.39) as

$$\hat{S}_{21}(\lambda) = \frac{B_2}{A_1} = \frac{V_2 - R_2 I_2}{V_1 + R_1 I_1}$$

$$= 2\frac{V_2}{V_g} \qquad (9.70)$$

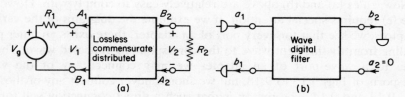

Fig. 9.26 (a) Reference filter, (b) wave digital equivalent.

Therefore S_{21} and $\hat{S}_{21}$ are identical except for a scaling factor. For many applications this is unimportant, however it can be compensated by adjusting the gain of either the A/D or D/A converter. Alternatively, an additional multiplier of appropriate value can be inserted at either the input or output. Henceforth, we shall not make any distinction between the general form of S_{21} and $\hat{S}_{21}$, remembering however that for the digital filter $S_{21} \neq S_{12}$ since the network is inherently non-reciprocal.

Before discussing some wave digital equivalents of commensurate distributed filters, let us summarize the procedure in general.

(i) A transfer function $S_{21}(\lambda)$ is derived which satisfies the required specifications regarding its amplitude and/or phase characteristics. This is the subject of a later chapter on approximation theory.

(ii) The transfer function is realized as a lossless commensurate distributed two-port with resistive terminations using the techniques of Chapter 8. This is the reference network. In this regard, we note that for many widely-used transfer characteristics, there exist explicit expressions and tables for the element values of the distributed filter. In particular, for transfer functions realizable as all-stub networks, the element values can be easily obtained from lumped prototypes for which extensive tables are available in the literature. However, this is not true of arbitrary transfer functions, and the reader is strongly reminded that the wave digital filter is generally derivable from a distributed reference network and not from a lumped prototype. The fact that lumped *ladder* prototypes can be used to obtain *all-stub* reference distributed networks is only a special case.

(iii) Having obtained the reference distributed network, the basic wave digital building blocks discussed in Section 9.7.2 are used to construct the wave digital equivalent, with the connections effected by adaptors.

When interconnecting the building blocks in the final step, a number of principles must be observed:

(a) The building blocks must be interconnected port by port, i.e. the wave-terminals of one wave-port must be connected to the two wave-terminals of exactly one other port.

(b) For every two wave-terminals which are connected together, the two corresponding waves must be compatible, i.e. they must flow in the same direction.

(c) The resulting wave flow diagram must be realizable.

Now, rules (a) and (b) above are relatively easy to comply with. However, rule (c) requires special attention. If we examine the diagrams of the various adaptors, we see that for every port of an adapter, there exists an inner path leading from the incident wave to the corresponding reflected wave. Therefore, to a wave-port of an adapter we may connect any of the wave one-ports of Figs 9.15 to 9.18, but we should never connect any of those in Figs 9.13 and 9.14. The reason is that such a direct connection will form a delay-free loop which is unrealizable in practice. For the same reason we must not connect the wave-port of an adapter to another adapter, but this may be connected to a UE (Fig. 9.17).

We now illustrate the synthesis of wave digital filters by considering some of the most commonly used types of distributed networks, namely: those employing UEs and/or stubs. The reader should note that all-stub distributed ladders, are the microwave counterparts of lumped LC ladders. However, a UE in the cascade mode has no lumped counterpart.

WAVE DIGITAL UE CASCADES

Consider the doubly terminated lossless commensurate distributed two-port shown in Fig. 9.27. It consists of three cascaded UEs of characteristic impedances Z_{01}, Z_{02}, Z_{03}. This third degree network is used to illustrate the technique of digitization, which naturally applies to an nth degree network.

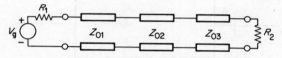

Fig. 9.27   Reference cascaded UEs filter.

Each UE has a wave digital equivalent of the type shown in Fig. 9.17. Thus cascade adaptors of the type shown in Fig. 9.25 can be used to construct the wave digital equivalent shown in Fig. 9.28. The values of the multipliers are obtained using (9.68); thus

$$\alpha_1 = \frac{R_1 - Z_{01}}{R_1 + Z_{01}} \tag{9.71a}$$

$$\alpha_2 = \frac{Z_{01} - Z_{02}}{Z_{01} + Z_{02}} \tag{9.71b}$$

$$\alpha_3 = \frac{Z_{02} - Z_{03}}{Z_{02} + Z_{03}} \tag{9.71c}$$

$$\alpha_4 = \frac{Z_{03} - R_2}{Z_{03} + R_2} \tag{9.71d}$$

However, the wave digital network shown in Fig. 9.28 is unrealizable due to the $T/2$ delays! This is because all the signals within the network have sample intervals of T. To overcome this difficulty, consider a basic section as shown in Fig. 9.29. If the inputs to this section were made to arrive $T/2$

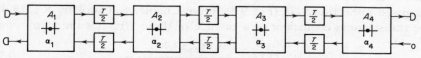

Fig. 9.28   Wave digital equivalent of the reference filter of Fig. 9.27.

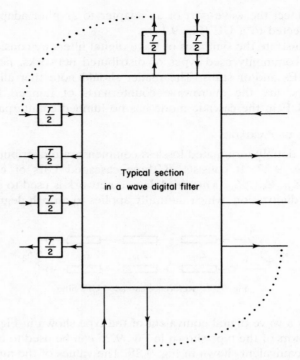

Fig. 9.29 Pertinent to the discussion of half-delays.

seconds earlier, and the outputs were delayed by $T/2$ seconds, then the new section of Fig. 9.30 would result. This can replace the original section. This technique can be used whenever $T/2$ delays are to be implemented.

Returning to the wave digital network of Fig. 9.28 we apply the above technique twice to arrive at the equivalent network of Fig. 9.31. First the technique is applied to adapter A_2 and the delays connected to it, i.e. the inputs to A_2 are entered $T/2$ earlier and the outputs are delayed by an extra $T/2$. Next the section containing adapter A_4 and the attached delays are considered; the inputs to A_4 are speeded up by $T/2$ and the outputs delayed by $T/2$. In this case, delays of $T/2$ and $-T/2$ appear at the reflected and incident waves of the output port. These, however, can be discarded for all practical purposes and have been omitted from Fig. 9.31.

At this stage, it is appropriate to compare this wave digital realization with an alternative direct realization of the same transfer function which was discussed in Section 9.51 (Fig. 9.5). From Chapter 8, Corollary 8.1, (8.35), the transmission coefficient of a cascade of three UEs has the form

$$S_{21}(\lambda) = \frac{(1-\lambda^2)^{3/2}}{a_0 + a_1\lambda + a_2\lambda^2 + a_3\lambda^3} \qquad (9.72)$$

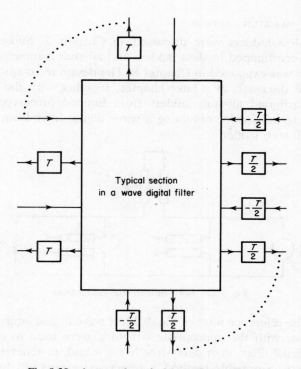

Fig. 9.30 A network equivalent to that of Fig. 9.29.

which is the general form of the transfer function realized by the network of Fig. 9.31. Now, if the transformation (9.19) is used to obtain

$$S_{21}(z) = H(z)$$
$$= \frac{z^{-1/2} \cdot z^{-1}}{b_0 + b_1 z^{-1} + b_2 z^{-2} + b_3 z^{-3}} \quad (9.73)$$

then (9.73) can be realized using a network of the form shown in Fig. 9.5, thus requiring four multipliers and three unit delays. The factor $z^{-1/2}$ in (9.73) can be extracted by the technique suggested for dealing with $T/2$ delays. The wave digital realization of Fig. 9.31 of the same function also requires four multipliers (one per adaptor) and three unit delays. Therefore Fig. 9.31 is also a canonic realization.

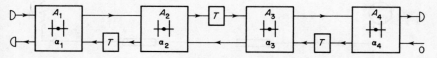

Fig. 9.31 Realizable wave digital equivalent of the UE cascade of Fig. 9.27.

WAVE DIGITAL ALL-STUB LADDERS

Lumped lossless ladders were discussed in Chapter 7. Subsequently the analogy between lumped lossless ladders and all-stub commensurate distributed ladders was explained in Chapter 8. The design techniques of lumped filters will be discussed in a later chapter, together with the methods of obtaining distributed all-stub ladders from lumped prototypes. We now consider the technique for obtaining a wave digital filter from a reference distributed all-stub ladder.

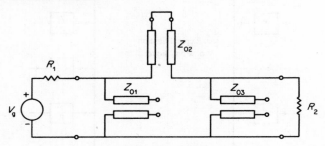

Fig. 9.32   Reference all-stub ladder filter.

Consider the reference filter of Fig. 9.32. If wave digital equivalents of the stubs, together with the appropriate adaptors, were used to construct the equivalent digital filter, then delay-free loops would, in general, be formed. To avoid these, a possible method is to separate the stubs by UEs using Kuroda's transformations of Chapter 8 (Section 8.5.2, Fig. 8.14). Then the wave digital equivalent is obtained in which the adaptors are separated by wave digital UEs. Figure 9.33 shows the stub–UE network after the application of Kuroda's transformations. The wave digital equivalent is shown in Fig. 9.34, and does not contain delay-free loops. So, once again, the UE comes to the rescue in circumventing a practical difficulty!

An alternative realization of the all-stub ladder is also possible without introducing UEs and avoids delay-free loops under certain conditions.[57] This is now discussed briefly.

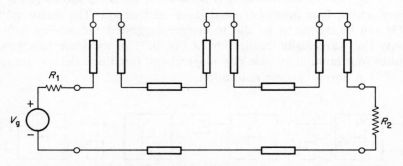

Fig. 9.33   The all-stub ladder of Fig. 9.32 after the application of Kuroda's transformations.

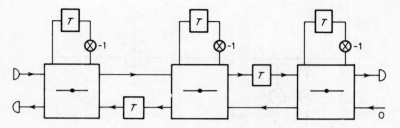

Fig. 9.34   Wave digital equivalent of Fig. 9.33.

Consider an n-port parallel adaptor of the type shown in Fig. 9.20. From (9.53) to (9.55) we can write the reflected wave at the nth port as

$$B_n = \sum_{k=1}^{n} \alpha_k A_k - A_n$$
$$= \sum_{k=1}^{n-1} \alpha_k A_k + (\alpha_n - 1) A_n \qquad (9.74)$$

Now, if we can choose

$$\alpha_n = 1 \qquad (9.75)$$

then B_n in (9.74) is independent of A_n, and this nth port can be connected directly to a port of another adaptor without forming a delay-free loop. From (9.55) it is clear that in order for the choice $\alpha_n = 1$ to be possible we must have

$$G_n = \sum_{k=1}^{n-1} G_k \qquad (9.76)$$

This particular port for which the above condition holds, is called the *reflection-free or matched port* and it is identified by a vertical stroke as shown in Fig. 9.35. Furthermore, since the multiplier value associated with the reflection-free port is unity, it can be omitted when calculating the number of multipliers needed for the realization. In addition to the matched port, we have seen that one of the other ports can be taken as the dependent port. Thus an n-port parallel adaptor with a matched port contains $(n-2)$ multipliers. Figure 9.36 shows a three-port parallel adaptor with port 3 as the matched port, and port 2 as the dependent port.

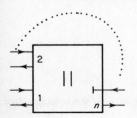

Fig. 9.35 n-Port parallel adaptor with port n as the reflection-free (matched) port.

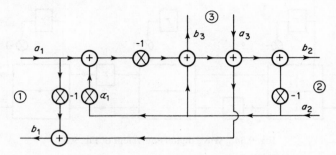

Fig. 9.36 Wave flow diagram of a three-port parallel adaptor with port 2 as the dependent port and port 3 as the reflection free (matched) port.

In a similar manner, we can obtain a series adaptor with one matched port. For the n-port series adapter shown in Fig. 9.22 we have from (9.60) to (9.62)

$$B_n = A_n - \beta_n \sum_{k=1}^{n} A_k$$

$$= A_n(1-\beta_n) - \beta_n \sum_{k=1}^{n-1} A_k \qquad (9.77)$$

and if we can choose

$$\beta_n = 1 \qquad (9.78)$$

then B_n is independent of A_n. Therefore, the possibility of a delay-free loop is eliminated. From (9.62) $\beta_n = 1$ requires

$$R_n = \sum_{k=1}^{n-1} R_k \qquad (9.79)$$

Again, one of the other ports can be chosen as the dependent port and we conclude that an n-port series adaptor with a matched port requires $(n-2)$ multipliers. This is shown in Fig. 9.37, where the stroke denotes the matched port.

Returning now to the all-stub reference ladder of Fig. 9.32, we can employ the types of adapters with matched ports to effect the connections

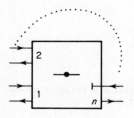

Fig. 9.37 n-Port series adaptor with port n as the reflection-free (matched) port.

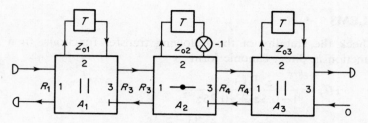

Fig. 9.38 Wave digital equivalent of the ladder of Fig. 9.32.

without forming delay-free loops. The resulting network is shown in Fig. 9.38. Adaptor A_1 is a three-port one with port 3 as the matched port, so that using (9.76) with $n = 3$ we take

$$\frac{1}{R_3} = \frac{1}{R_1} + \frac{1}{Z_{01}} \tag{9.80}$$

Similarly A_2 is a series three-port adaptor with its port 3 as the matched port so that using (9.79) we choose

$$R_4 = R_3 + Z_{02} \tag{9.81}$$

Finally, we note that the wave digital network of Fig. 9.38 has the true ladder topology and it is the direct equivalent of the all-stub ladder of Fig. 9.32.

9.8 CONCLUSION

The synthesis techniques of digital transfer functions were discussed in some detail. First, techniques peculiar to digital filters were given and it was shown that simple direct realizations are possible for both IIR and FIR filters. However, the direct recursive realizations of IIR filters suffer from a high sensitivity problem, and possible solutions were pointed out. The first consists in factoring the transfer function into first and second order terms, realizing them separately, then connecting the resulting subnetworks in cascade. An alternative realization proceeds from the partial fraction expansion of the transfer function and leads to the parallel form.

With the sensitivity problem in mind, wave digital filters were discussed. These imitate the low sensitivity properties of the classical passive structures discussed in earlier chapters. We have seen that the synthesis techniques of commensurate distributed networks can be applied to obtain practical wave digital filters with the same transfer functions as those of reference passive networks. We have only discussed the wave digital equivalents of the more commonly used types of commensurate distributed networks. However, it is also possible to obtain wave digital equivalents of many coupled-line distributed networks; the interested reader may consult the references.[58–60]

PROBLEMS

9.1 Check the stability of the following transfer functions, then realize each function in direct canonic form

(a) $$H(z) = \frac{z^{-1}(1+z^{-1})}{1-0.5z^{-1}+0.25z^{-2}}$$

(b) $$H(z) = \frac{z^{-1}(1-z^{-1})^3}{1+1.75z^{-1}+0.5z^{-2}+0.25z^{-3}+0.25z^{-4}}$$

9.2 Find the non-recursive direct realization of the FIR transfer function

$$H(z) = 1 + z^{-1} + z^{-2} + 3z^{-3} + 5z^{-4} + 2z^{-5}$$

9.3 Realize the following IIR transfer function once in cascade form, and a second time in parallel form

$$H(z) = \frac{z^{-1}(1+z^{-1})}{(1+0.75z^{-1})(1-0.5z^{-1}+0.25z^{-2})}$$

9.4 Obtain the wave digital realization of the following transfer functions

(a) $$S_{21}(\lambda) = \frac{(1-\lambda^2)^{3/2}}{6+11\lambda+6\lambda^2+\lambda^3}$$

(b) $$S_{21}(\lambda) = \frac{1}{\lambda^2+5\lambda+6}$$

both with 1 Ω terminating resistors.

9.5 By first transforming the transfer functions in Problem 9.4 into the z^{-1}-domain, realize each function in direct and cascade forms.

9.6 Obtain the wave digital equivalent of the doubly-terminated cascade of UEs shown in Fig. P9.6. Calculate the transfer function of the filter and sketch its magnitude versus frequency for a sampling frequency of 20 kHz.

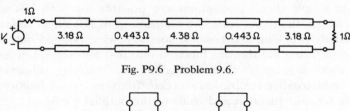

Fig. P9.6   Problem 9.6.

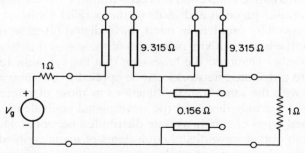

Fig. P9.7   Problem 9.7.

9.7 Using the low-pass all-stub ladder filter shown in Fig. P9.7 as a reference filter, obtain a wave digital filter possessing the same transfer function, and having the true ladder topology. Sketch the magnitude of the filter transfer function against ω/ω_N, where ω_N is the radian sampling frequency.

PART IV

Approximation Theory and Filter Design

OUTLINE

So far in this book we have considered the synthesis techniques of a given function (or matrix) in terms of a well-defined set of circuit elements. In this, the final part, we discuss the problem of the derivation of the required function such that the realized network performs a certain task. In particular, our interest is concentrated in transfer functions of filters. This is the approximation problem in filter design, which may be defined, more succinctly, as that of obtaining a realizable transfer function which produces characteristics meeting the required specifications regarding its amplitude and/or phase.

This part is divided into four chapters. The first three (Chapters 10–12) deal with lumped, commensurate distributed, and digital filters, respectively. Chapter 13 gives a brief outline of the design technique of a class of switched-capacitor filters which are modelled on passive prototypes.

PART-IV

Approximation Theory
and Filter Design

OUTLINE

Chapter 10
Approximation Methods for Lumped Filters

10.1 INTRODUCTION

10.1.1 Statement of the problem

The approximation problem for passive lumped filters is discussed in this chapter. The required filter is a lossless reciprocal two-port N to operate between a resistive source and a resistive load as shown in Fig. 10.1. We also assume that the source resistor has a value of $1\,\Omega$, but later in the synthesis all impedance values of the realized network can be scaled to accommodate the actual termination.

The lossless two-port N is characterized by its scattering matrix $[S(p)]$ normalized to r_g ($=1\,\Omega$) and r_e. The approximation problem consists in the derivation of a realizable transfer function $S_{21}(p)$ with the required properties. From Section 6.3.2 $S_{21}(p)$ represents the scattering transmission coefficient of a passive lossless two-port if and only if it is a bounded real function, i.e. in addition to being analytic in the closed right half-plane (Re $p \geq 0$) it must satisfy

$$|S_{21}(j\omega)| \leq 1 \qquad -\infty \leq \omega \leq \infty \tag{10.1}$$

Writing $S_{21}(j\omega)$ as

$$S_{21}(j\omega) = |S_{21}(j\omega)|\, e^{j\psi(\omega)} \tag{10.2}$$

then from (5.26)

$$|S_{21}(j\omega)|^2 = G(\omega^2) \tag{10.3}$$

which is the transducer power gain, i.e. the ratio of the power delivered to the load, to that available from the source. In (10.2) $\psi(\omega)$ is the phase function.

The transfer function $S_{21}(p)$ may be written as

$$S_{21}(p) = \frac{f_m(p)}{g_n(p)} \qquad m \leq n \tag{10.4}$$

where the condition $m \leq n$ must be satisfied since S_{21} is analytic along the

199

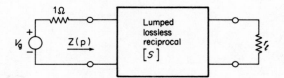

Fig. 10.1   Normalized filter.

entire $j\omega$-axis including the point at infinity. Of course $g_n(p)$ is a strictly Hurwitz polynomial, and if the network is to be reciprocal f_m must be either even or odd so that $S_{12} = S_{21}$.

Now, the ideal amplitude characteristics for low-pass, high-pass, band-pass, and band-stop filters are shown in Fig. 10.2. These cannot be attained using realizable transfer functions and must, therefore, be approximated. Similarly, the ideal (no distortion) phase characteristic $\psi(\omega)$ is a linear function of frequency in the passband, e.g. for the low-pass case,

$$-\psi(\omega) = k\omega \qquad \omega \leq \omega_0 \tag{10.5}$$

as shown in Fig. 10.3. Again, this must be approximated by realizable transfer functions.

In some applications, such as voice communication, filters are designed on amplitude basis only since the human ear is relatively insensitive to phase distortion. Other applications tolerate some amplitude distortion while requiring a close approximation to the ideal (linear) phase response. However, in modern high capacity communication systems, filters are required to possess good amplitude (highly selective) as well as good phase characteristics.

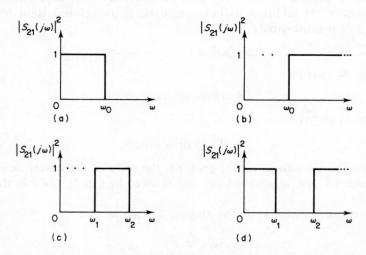

Fig. 10.2   Ideal amplitude responses. (a) Low-pass, (b) high-pass, (c) band-pass, (d) band-stop.

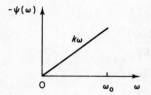

Fig. 10.3   Ideal low-pass phase response.

In this chapter, the problems of amplitude and phase approximations are donsidered separately at first. Then combined amplitude and phase approximations are discussed. Once the required realizable transfer function $S_{21}(p)$ (or $|S_{21}(j\omega)|^2$) has been determined, the appropriate synthesis techniques of Part II are employed to construct the filter. Thus, the input impedance of the resistor-terminated lossless two-port of Fig. 10.1 is determined according to Steps (1) to (5), expressions (6.22) to (6.27) in Section 6.3.2. Then the impedance is realized using the techniques of Chapters 6 and 7. We note, in passing, that some commonly used filter transfer functions lead to explicit formulae for the element values. These will be given wherever possible together with the references where tables for element values may be found.

10.1.2   Amplitude and phase functions

The transmission loss function (or attenuation function) of the two-port in Fig. 10.1 is given by

$$\alpha(\omega) = 10 \log \frac{1}{|S_{21}(j\omega)|^2} \tag{10.6}$$

From (5.29) we have

$$|S_{11}(j\omega)|^2 + |S_{21}(j\omega)|^2 = 1 \tag{10.7}$$

and the return (or echo) loss is given by

$$\alpha_R(\omega) = 10 \log \frac{1}{|S_{11}(j\omega)|^2}$$

$$= 10 \log \frac{1}{1 - |S_{21}(j\omega)|^2} \tag{10.8}$$

As we shall see shortly, the specifications on the amplitude response of the filter are often given in terms of the stopband transmission loss and passband return loss.

In the treatment of phase approximation, the problem may be stated directly in terms of the required phase $\psi(\omega)$. Alternatively, in fact very often, the problem can be formulated in terms of either the group-delay

$T_g(\omega)$ or phase delay $T_{ph}(\omega)$ defined by

$$T_g(\omega) = -\frac{d\psi(\omega)}{d\omega} \tag{10.9}$$

$$T_{ph}(\omega) = -\frac{\psi(\omega)}{\omega} \tag{10.10}$$

For an approximation to the ideal (no distortion) phase characteristics $\psi(\omega)$ is required to approximate a linear function of ω at all frequencies in the passband. Alternatively, the group-delay in (10.9) is required to approximate to a constant within the passband.

Now, from (10.2) and (10.9) we have

$$T_g(\omega) = -\text{Ev}\left\{\frac{d}{dp}\ln S_{21}(p)\right\}\Big|_{p=j\omega} \tag{10.11}$$

Letting

$$T(p) = -\frac{d}{dp}\ln S_{21}(p) \tag{10.12}$$

and using (10.4) we may write

$$T(p) = \frac{g'}{g} - \frac{f'}{f} \tag{10.13}$$

where the prime denotes differentiation. Thus

$$T_g(p) = \text{Ev}\, T(p)$$
$$= \frac{1}{2}\left(\frac{g'}{g} + \frac{g'_*}{g_*} - \frac{f'}{f} - \frac{f'_*}{f_*}\right) \tag{10.14}$$

where, as usual the lower asterisk denotes replacing p by $-p$. $T_g(\omega)$ is obtained by evaluating (10.14) at $p = j\omega$.

Let us now examine the question of the relationship between the amplitude and phase of a function. Consider a special type of transfer function which has no zeros in the open right half-plane, i.e. in (10.4) S_{21} is such that

$$f_m(p) \neq 0 \quad \text{in Re } p > 0 \tag{10.15}$$

In such a case S_{21} is termed a *minimum-phase* function. This is because under the constraint (10.15) the phase shift $\psi(\omega)$ of $S_{21}(j\omega)$ is a minimum over the range $-\infty \leq \omega \leq \infty$ for a given transducer power gain $|S_{21}(j\omega)|^2$. For such a function, there exists a unique relationship between the amplitude and phase responses at real frequencies ($p = j\omega$). This can be obtained by considering the functions

$$\frac{\ln S_{21}(p)}{(p^2 + \omega_0^2)} \quad \text{and} \quad \frac{\ln S_{21}(p)}{p(p^2 + \omega_0^2)} \tag{10.16}$$

which are analytic in Re $p>0$. Letting

$$S_{21}(j\omega) = \exp\{-\gamma(\omega) + j\psi(\omega)\} \tag{10.17}$$

and performing a contour integration enclosing the right half-plane we have

$$\psi(\omega_0) = -\frac{\omega_0}{\pi} \int_{-\infty}^{\infty} \frac{\gamma(\omega)}{\omega^2 - \omega_0^2} \, d\omega \tag{10.18}$$

$$\gamma(\omega_0) = \gamma(0) + \frac{\omega_0^2}{\pi} \int_{-\infty}^{\infty} \frac{\psi(\omega)}{\omega(\omega^2 - \omega_0^2)} \, d\omega \tag{10.19}$$

i.e. $\psi(\omega)$ and $\gamma(\omega)$ form a Hilbert transform pair. Thus, at a particular frequency ω_0, the phase delay $-\psi(\omega_0)/\omega_0$ is mainly determined by the behaviour of the amplitude response in the vicinity of ω_0 since the integrand in (10.18) becomes very large in that region. It follows that as the rate of cutoff increases, the phase delay varies more rapidly. Hence, a minimum-phase transfer function with good amplitude response is usually associated with a poor phase characteristic. For this reason, non-minimum-phase functions must be used to achieve a design combining high selectivity with a good approximation to phase linearity.

Having disposed of the above preliminaries, we now turn to the approximation problem itself. At first, amplitude and phase approximations are treated separately, then the combined amplitude and phase approximation problem is considered.

10.2 AMPLITUDE APPROXIMATION

10.2.1 Format of specifications

The transducer power gain of a real finite lossless two-port may be written as

$$|S_{21}(j\omega)|^2 = G(\omega^2)$$

$$= \frac{\displaystyle\sum_{r=0}^{m} a_r \omega^{2r}}{\displaystyle\sum_{r=0}^{n} b_r \omega^{2r}} \tag{10.20}$$

which may be put in the form

$$|S_{21}(j\omega)|^2 = G(\omega^2)$$

$$= \frac{1}{1 + X_n(\omega^2)} \tag{10.21}$$

where $X_n(\omega^2)$ is a real rational function of ω^2 which is non-negative for all real values of ω,

$$X_n(\omega^2) \geq 0 \qquad -\infty \leq \omega \leq \infty \tag{10.22}$$

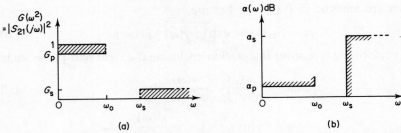

Fig. 10.4 Format of specifications on the amplitude response of a low-pass filter. (a) Transducer power gain, (b) attenuation (transmission loss).

We shall begin by considering the design of low-pass filters, then discuss the methods of obtaining other types. In the low-pass case, amplitude approximation consists in determining $|S_{21}(j\omega)|^2$ such that it meets the typical specifications shown in Fig. 10.4. This is a tolerance scheme which has the following general features

(i) In the passband

$$G_p \le G(\omega^2) \le 1 \qquad 0 \le \omega \le \omega_0 \qquad (10.23a)$$

or in terms of the transmission loss (attenuation)

$$\alpha(\omega) \le \alpha_p \qquad 0 \le \omega \le \omega_0 \qquad (10.23b)$$

where

$$\alpha_p = 10 \log G_p^{-1} \, dB \qquad (10.23c)$$

(ii) In the stopband

$$0 \le G(\omega^2) \le G_s \qquad \omega \ge \omega_s \qquad (10.24a)$$

or in terms of the transmission loss (attenuation)

$$\alpha(\omega) \ge \alpha_s \qquad \omega \ge \omega_s \qquad (10.24b)$$

where

$$\alpha_s = 10 \log G_s^{-1} \, dB \qquad (10.24c)$$

(iii) In the transition band, the power gain is assumed to decrease monotonically from G_p at ω_0 to G_s at ω_s. Thus the transmission loss (or attenuation) $\alpha(\omega)$ increases monotonically from α_p at ω_0 to α_s at ω_s. The rate of cutoff (or selectivity) is measured by the ratio of stopband edge ω_s to passband edge ω_0,

$$\gamma = \omega_s/\omega_0 \qquad (10.25)$$

(iv) At ω_0, $G(\omega^2)$ assumes its smallest passband value G_p, and at ω_s its largest stopband value G_s.

We now discuss the most commonly used types of solutions to the amplitude approximation problem.

10.2.2 Maximally flat response in both passband and stopband

This is the so-called Butterworth filter; the general appearance of its response is shown in Fig. 10.5. In this case the maximum number of derivatives of $|S_{21}(j\omega)|^2$ with respect to ω, is made to vanish at both $\omega = 0$ and $\omega = \infty$. First, the condition $|S_{21}(0)| = 1$ and $|S_{21}(\infty)| = 0$ lead to $|S_{21}|^2$ in (10.20) being of the form

$$|S_{21}(j\omega)|^2 = \frac{1 + \sum_{r=1}^{n-1} a_r \omega^{2r}}{1 + \sum_{r=1}^{n} b_r \omega^{2r}} \tag{10.26}$$

The maximally flat condition at the origin requires that the first $(2n-1)$ derivatives of $|S_{21}|^2$ be equated to zero. To apply this condition we use (10.26) to write

$$1 - |S_{21}(j\omega)|^2 = \frac{\sum_{r=1}^{n-1} (b_r - a_r)\omega^{2r} + b_n \omega^{2n}}{1 + \sum_{r=1}^{n} b_r \omega^{2r}} \tag{10.27}$$

and the above function together with its first $(2n-1)$ derivatives are required to vanish at $\omega = 0$. Hence, the power series expansion of (10.27) around $\omega = 0$ must be of the form

$$1 - |S_{21}(j\omega)|^2 = c_n \omega^{2n} + c_{n+1}\omega^{2n+2} + \ldots \tag{10.28}$$

which implies

$$a_r = b_r \qquad r = 1, 2, \ldots, (n-1) \tag{10.29}$$

To apply the maximally flat condition at $\omega = \infty$ we rewrite (10.26) as

$$|S_{21}(j\omega)|^2 = \frac{\sum_{r=1}^{n-1} a_r \omega^{2r-2n} + \omega^{-2n}}{\sum_{r=1}^{n} b_r \omega^{2r-2n} + \omega^{-2n}} \tag{10.30}$$

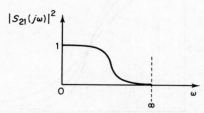

Fig. 10.5 Maximally flat amplitude response in both passband and stopband (Butterworth).

and the series expansion of (10.30) around $\omega = \infty$ is required to be of the form

$$|S_{21}(j\omega)|^2 = d_n\omega^{-2n} + d_{n+1}\omega^{-(2n+2)} + \ldots \tag{10.31}$$

which gives

$$a_r = 0 \qquad r = 1, 2, \ldots, (n-1) \tag{10.32}$$

Combining (10.29) with (10.32), $|S_{21}(j\omega)|^2$ in (10.26) takes the form

$$|S_{21}(j\omega)|^2 = \frac{1}{1 + b_n\omega^{2n}} \tag{10.33}$$

But since ω can be scaled by any arbitrary constant without affecting the maximally flat nature of the response, this constant is normally chosen such that in (10.33) $b_n = 1$. Thus

$$|S_{21}(j\omega)|^2 = \frac{1}{1 + \omega^{2n}} \tag{10.34}$$

with the 3-dB occurring at $\omega = 1$ for all n. Later in the synthesis we may use frequency scaling to transform the 3-dB point to any arbitrary value of ω. Typical plots of (10.34) are shown in Fig. 10.6, which shows that all aspects of the response improve with increasing the degree of the function.

To obtain $S_{21}(p)$ from (10.40) we first find the poles of $|S_{21}(j\omega)|^2$. These occur at

$$\omega^{2n} = -1 = \exp\{j(2r-1)\pi\}, \qquad r = 1, 2, \ldots, 2n$$

i.e. at

$$\omega = \exp\left\{\frac{j(2r-1)\pi}{2n}\right\}$$

or

$$p = j \exp j\theta_r$$
$$= -\sin\theta_r + j\cos\theta_r \tag{10.35}$$

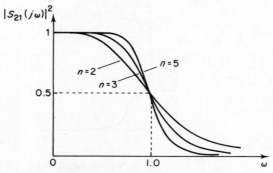

Fig. 10.6 Typical responses of the maximally flat filter described by (10.34).

where

$$\theta_r = \frac{(2r-1)}{2n}\,\pi \tag{10.36}$$

Left half-plane poles are selected for a strictly Hurwitz denominator and the resulting bounded real transfer function is given by

$$S_{21}(p) = \frac{1}{\prod\limits_{r=1}^{n}\{p - j\exp(j\theta_r)\}} \tag{10.37}$$

where

$$\theta_r = \frac{(2r-1)}{2n}\,\pi \qquad r = 1, 2, \ldots, n \tag{10.38}$$

Now since $S_{21}(p)$ is an all-pole transfer function, all the transmission zeros are at $p = \infty$ and $S_{21}(p)$ satisfies (7.5). Therefore the function is realizable as a simple LC ladder using the technique of Section 7.2. Thus,

$$|S_{11}(j\omega)|^2 = 1 - |S_{21}(j\omega)|^2$$

$$= \frac{\omega^{2n}}{1 + \omega^{2n}} \tag{10.39}$$

Then letting $\omega \to p/j$ we obtain

$$S_{11}S_{11*} = \frac{(-1)^n p^{2n}}{1 + (-p)^{2n}} \tag{10.40}$$

and noting that the poles of S_{11} are the same as those of S_{21} we have

$$S_{11}(p) = \frac{\pm p^n}{\prod\limits_{l}\{p - j\exp(j\theta_r)\}} \tag{10.41}$$

The input impedance of the resistor terminated ladder is obtained from (6.26) as

$$Z(p) = \frac{1 + S_{11}(p)}{1 - S_{11}(p)} \tag{10.42}$$

If the positive sign in (10.41) is chosen, then $Z(p)$ will have a pole at $p = \infty$ and the network of Fig. 10.7(a) results. On the other hand, if the negative sign in (10.41) is chosen, then $Y(p) = Z^{-1}(p)$ will have a pole at $p = \infty$ and the network of Fig. 10.7(b) results. For both networks, there exist explicit formulae for the element values in terms of θ_r in (10.38). These are given by

$$g_r = 2\sin\theta_r \qquad r = 1, 2, \ldots, n \tag{10.43}$$

where

$$g_r = L_r \quad \text{or} \quad C_r \tag{10.44}$$

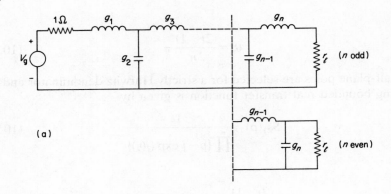

(a)

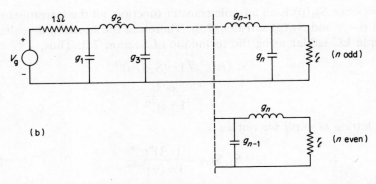

(b)

Fig. 10.7 Alternative realizations of maximally flat and Chebyshev low-pass prototype filters.

and

$$r_\ell = 1 \ \Omega \tag{10.45}$$

Note that the above formulae give the element values of the low-pass equally terminated prototype ($r_g = r_\ell = 1 \ \Omega$) of Fig. 10.7 with the 3-dB point at $\omega = 1$. It is also clear that since the numerator of S_{11} in (10.41) is either even or odd, then the resulting network is either symmetric (n odd) or antimetric (n even), as explained in Section 5.7.3. Naturally, this is also evident from (10.43).

Now the degree of the required filter can be obtained from the given set of specifications. These may be given, or re-expressed, in either of the two following forms.

(a) 3-dB point at $\omega = 1$
Stopband edge at $\omega = \omega_s$ with $\alpha(\omega) \geq \alpha_s$ for $\omega \geq \omega_s$.
From (10.34) we require

$$10 \log (1 + \omega_s^{2n}) \geq \alpha_s$$

which gives

$$n \geq \frac{\log (10^{0.1\alpha_s} - 1)}{2 \log \omega_s} \qquad (10.46)$$

in which ω_s is the actual frequency *normalized* with respect to the 3-dB point. Normally α_s will be sufficiently large to enable unity to be neglected with respect to $10^{0.1\alpha_s}$. Thus,

$$n \geq \frac{\alpha_s}{20 \log \omega_s} \qquad (10.47)$$

(b) Alternatively, the specifications may be given (or re-expressed) in the following format.

Minimum stopband attenuation $= \alpha_s$, $\omega \geq \omega_s$
Minimum passband return loss $= \alpha_{Rp}$, $\omega \leq \omega_0$
Ratio of stopband to passband edges $\omega_s/\omega_0 = \gamma$

Here, the passband edge is defined at a frequency which is not necessarily the 3-dB point. In this case we require in the passband

$$10 \log \frac{1}{|S_{11R}|} \geq \alpha_{Rp}$$

and use of (10.40) in (10.8) gives

$$10 \log (1 + \omega_0^{-2n}) \geq \alpha_{Rp}$$

If α_{Rp} is sufficiently large we have

$$n \geq \frac{\alpha_{Rp}}{20 \log \omega_0^{-1}} \qquad (10.48)$$

For the stopband (10.47) still holds, which is combined with (10.48) to give for the required degree

$$n \geq \frac{\alpha_s + \alpha_{Rp}}{20 \log \gamma}, \qquad \gamma = \frac{\omega_s}{\omega_0} \qquad (10.49)$$

10.2.3 Chebyshev response

In this type of approximation, $|S_{21}(j\omega)|^2$ is required to be equiripple within the passband and have a maximally flat response in the stopband, as shown in the typical response of Fig. 10.8. For a low-pass prototype with passband edge at $\omega = 1$, we use (10.32) to force the function to have $(2n - 1)$ zero derivatives at $\omega = \infty$. We may then write

$$|S_{21}(j\omega)|^2 = \frac{1}{1 + \varepsilon^2 T_n^2(\omega)} \qquad (10.50)$$

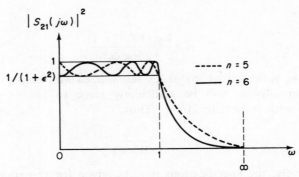

Fig. 10.8 Typical responses of low-pass Chebyshev prototype filters.

where $T_n(\omega)$ is chosen as an odd or even polynomial which oscillates
between -1 and $+1$ the maximum number of times in the interval $|\omega| \leq 1$
and is monotonic outside this interval. The required behaviour is shown in
Fig. 10.9. Consequently $|S_{21}(j\omega)|^2$ oscillates between 1 and $1/(1 + \varepsilon^2)$ in the
interval $|\omega| \leq 1$. The size of the ripple can be determined by a suitable choice
of ε. All points in the passband $|\omega| \leq 1$, where $|T_n(\omega)| = 1$ must be maxima or
minima. Thus

$$\frac{dT_n(\omega)}{d\omega}\Bigg|_{|T_n(\omega)|=1} = 0 \qquad \text{except where } |\omega| = 1 \qquad (10.51)$$

Hence the required polynomial satisfies the differential equation

$$\frac{dT_n(\omega)}{d\omega} = K\frac{\sqrt{1 - T_n^2(\omega)}}{\sqrt{1 - \omega^2}} \qquad (10.52)$$

or

$$\frac{dT_n(\omega)}{\sqrt{1 - T_n^2(\omega)}} = K\frac{d\omega}{\sqrt{1 - \omega^2}} \qquad (10.53)$$

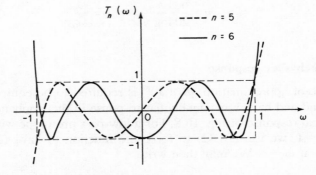

Fig. 10.9 Typical plots of the Chebyshev polynomial of the
first kind.

where K is a constant to be determined from the condition that $T_n(\omega)$ is a polynomial of exact degree n. Integrating (10.53) we obtain

$$\cos^{-1} T_n(\omega) = K \cos^{-1} \omega$$

or

$$T_n(\omega) = \cos (K \cos^{-1} \omega) \tag{10.54}$$

and if we put $K = n$, then $T_n(\omega)$ is an nth degree polynomial. Hence,

$$T_n(\omega) = \cos (n \cos^{-1} \omega) \qquad 0 \le |\omega| \le 1 \tag{10.55a}$$

$$= \cosh (n \cosh^{-1} \omega) \qquad |\omega| > 1 \tag{10.55b}$$

which is the Chebyshev polynomial of the first kind. $T_n(\omega)$ can be easily generated using the recurrence formula

$$T_{n+1}(\omega) = 2\omega T_n(\omega) - T_{n-1}(\omega) \tag{10.56}$$

with

$$T_0(\omega) = 1, \quad T_1(\omega) = \omega$$

Examples are

$$T_2(\omega) = 2\omega^2 - 1$$
$$T_3(\omega) = 4\omega^3 - 3\omega$$
$$T_4(\omega) = 8\omega^4 - 8\omega^2 + 1$$
$$T_5(\omega) = 16\omega^5 - 20\omega^3 + 5\omega \tag{10.57}$$

We also note that for n odd $T_n(0) = 0$ and $|S_{21}(0)| = 1$, whereas for n even $T_n(0) = 1$ and $|S_{21}(0)|^2 = 1/(1 + \varepsilon^2)$. In general the parameter ε in (10.50) controls the passband ripple and can be chosen freely. On the other hand, increasing n improves the stopband attenuation since the value of the polynomial $T_n(\omega)$ increases with increasing n for $\omega > 1$. A combination of n and ε can always be found to meet any arbitrary set of specifications, as will be seen shortly.

It can be shown, either by a rigorous proof, or by a simple heuristic argument,[8] that the Chebyshev approximation is the optimum solution to the problem of determining an $|S_{21}|^2$ which is constrained to lie in a band for $0 \le \omega \le 1$ and attain a minimum value for all ω in the range $\omega_1 \le \omega \le \infty$ with $\omega_1 > 1$, for a given degree.

Now the synthesis of the Chebyshev transfer function can be accomplished in the usual manner. First note that the transfer function $S_{21}(p)$ which would be obtained from (10.50) is, again, an all-pole function. Therefore, the Chebyshev function is realizable as a ladder of the same form shown in Fig. 10.7. As usual, we form

$$|S_{11}(j\omega)|^2 = 1 - |S_{21}(j\omega)|^2$$

$$= \frac{\varepsilon^2 T_n^2(\omega)}{1 + \varepsilon^2 T_n^2(\omega)} \tag{10.58}$$

and

$$S_{11}S_{11*} = \frac{\varepsilon^2 T_n^2(p/j)}{1 + \varepsilon^2 T_n^2(p/j)} \tag{10.59}$$

Then the above expression is factored to form $S_{11}(p)$. The zeros of $S_{11}S_{11*}$ occur at

$$T_n^2(p/j) = \cos^2(n \cos^{-1} p/j) = 0 \tag{10.60}$$

or

$$p = j \cos \theta_r \tag{10.61}$$

with

$$\theta_r = \frac{(2r-1)}{2n} \pi \tag{10.62}$$

The poles of (10.59) occur at

$$\varepsilon^2 T_n^2(\omega) = -1 \tag{10.63}$$

Defining an auxiliary parameter as

$$\eta = \sinh\left(\frac{1}{n} \sinh^{-1} \frac{1}{\varepsilon}\right) \tag{10.64}$$

then, the poles occur at

$$\cos^2(n \cos^{-1} \omega) = -\sinh^2(n \sinh^{-1} \eta) \tag{10.65}$$

or

$$n \cos^{-1} \omega = n \sin^{-1} j\eta + (2r-1)\pi/2 \tag{10.66}$$

i.e.

$$p = -j \cos\{\sin^{-1} j\eta + \theta_r\} \tag{10.67}$$

where θ_r is given by (10.62). For a bounded real $S_{11}(p)$ we select the open left half-plane poles. This gives for the denominator of $S_{11}(p)$

$$\prod_{r=1}^{n} \{p + j \cos(\sin^{-1} j\eta + \theta_r)\} \tag{10.68}$$

Regarding the selection of the zeros of $S_{11}(p)$ one may take any convenient set of those given by (10.61), with no constraint except that the factors must combine to form an nth degree real polynomial. For example, we can choose all the left half-plane zeros, or we may take those alternating between the two half-planes. Each choice leads to a network with different element values but both have the same transducer power gain.

For a minimum-phase reflection coefficient $S_{11}(p)$ all the zeros lie in the left half-plane and for this choice we have from (10.61) to (10.68)

$$S_{11}(p) = \pm \prod_{r=1}^{n} \left\{ \frac{p + j \cos \theta_r}{p + (\eta \sin \theta_r + j\sqrt{1 + \eta^2} \cos \theta_r)} \right\} \tag{10.69}$$

Noting that the poles of $S_{11}(p)$ are those of $S_{21}(p)$ we also have

$$S_{21}(p) = \frac{\prod_{r=1}^{n} \left\{ \eta^2 + \sin^2 \left(\frac{r\pi}{n} \right) \right\}^{1/2}}{\prod_{r=1}^{n} \left\{ p + (\eta \sin \theta_r + j\sqrt{1+\eta^2} \cos \theta_r) \right\}} \qquad (10.70)$$

where the constant factor in the numerator is such that for n odd $S_{21}(0) = 1$ and for n even $S_{21}(0) = 1/\sqrt{1+\varepsilon^2}$.

We also note that the numerator of $S_{11}(p)$ is either odd or even leading to a symmetric (for n odd) or antimetric (for n even) ladder, as explained in Section 5.7.3. Having formed $S_{11}(p)$ using (10.69), the synthesis is accomplished by calculating $Z(p)$ from (10.42) and employing the technique of Section 7.2. In fact, the element values in the networks of Fig. 10.7 for the Chebyshev case are given by

$$g_1 = \frac{2}{\eta} \sin \left(\frac{\pi}{2n} \right) \qquad (10.71a)$$

$$g_r g_{r+1} = \frac{4 \sin \theta_r \sin \left(\frac{2r+1}{2n} \right) \pi}{\eta^2 + \sin^2 \left(\frac{r\pi}{n} \right)} \qquad (10.71b)$$

with

$$g_r = L_r \quad \text{or} \quad C_r \qquad (10.71c)$$

and the passband edge at $\omega = 1$. For n odd, $S_{21}(0) = 1$ and the load resistor $r_\ell = r_g = 1 \, \Omega$. However, for n even $|S_{21}(0)|^2 = 1/(1+\varepsilon^2)$ and $r_\ell \neq 1$. At $\omega = 0$ all inductors become short circuits and capacitors become open circuits. Therefore for n even

$$|S_{21}(0)|^2 = \frac{4r_\ell}{(r_\ell + 1)^2} \qquad (10.72)$$

and r_ℓ is determined in terms of ε as

(a) $\qquad\qquad r_\ell = (\varepsilon + \sqrt{1+\varepsilon^2})^2 \qquad \text{for } S_{11}(0) > 0 \qquad (10.73a)$

(b) $\qquad\qquad r_\ell = (\varepsilon + \sqrt{1+\varepsilon^2})^{-2} \qquad \text{for } S_{11}(0) < 0 \qquad (10.73b)$

Finally consider the degree of the Chebyshev filter required to meet a typical set of specifications. Let these be expressed as follows.

Passband attenuation (transmission loss)

$$\alpha(\omega) \leq \alpha_p \qquad 0 \leq \omega \leq 1$$

Stopband attenuation (transmission loss)

$$\alpha(\omega) \geq \alpha_s \qquad \omega \geq \omega_s$$

Then from (10.50) we require in the passband

$$10 \log (1 + \varepsilon^2) \leq \alpha_p$$

i.e.

$$\varepsilon^2 \leq 10^{0.1\alpha_p} - 1 \tag{10.74}$$

At the stopband edge we require

$$10 \log \{1 + [\varepsilon \cosh (n \cosh^{-1} \omega_s)]^2\} \geq \alpha_s \tag{10.75}$$

Solving for n and using (10.74) we have for the required degree

$$n \geq \frac{\cosh^{-1} \{(10^{0.1\alpha_s} - 1)/(10^{0.1\alpha_p} - 1)\}^{1/2}}{\cosh^{-1} \omega_s} \tag{10.76}$$

Alternatively, the specifications may be given or re-expressed in terms of the stopband edge ω_s (normalized to the passband edge), minimum passband return loss α_{Rp}, and minimum stopband attenuation α_s. Then, an approximate design formula gives the degree of the required filter as

$$n \geq \frac{\alpha_s + \alpha_{Rp} + 6}{20 \log (\omega_s + \sqrt{\omega_s^2 - 1})} \tag{10.77}$$

10.2.4 Elliptic function response

This is the case of equiripple response in both the passband and stopband. It is well-known[8] that this equiripple solution is optimum in the sense of minimizing the maximum deviation in each band of a rational function, for the two-band approximation shown in Fig. 10.4(a), for a given degree. We have seen that, in general, the power gain of the lossless filter is expressible as in (10.21). In the maximally flat and Chebyshev types of approximation $X(\omega^2)$ reduces to the square of a polynomial. So, a more general type of approximation can be obtained if we allow $X(\omega^2)$ to be the square of a rational function.

Consider Fig. 10.10 which shows a typical low-pass response of the optimum equiripple prototype. Write

$$|S_{21}(j\omega)|^2 = \frac{1}{1 + \varepsilon^2 F_n^2(\omega)} \tag{10.78}$$

The general behaviour of the required function $F_n(\omega)$ is shown in Fig. 10.11. The main features are as follows.[8]

(a) $F_n(\omega)$ has all its n zeros in the interval $|\omega| < 1$ and all its poles outside this interval.
(b) $F_n(\omega)$ oscillates between the values ± 1 in the interval $|\omega| < 1$
(c) $F_n(1) = 1$
(d) $1/F_n(\omega)$ oscillates between $\pm 1/M$ in the interval $|\omega| > \omega_s$, where

$$M^2 = \frac{G_s^{-1} - 1}{G_p^{-1} - 1} = \frac{10^{0.1\alpha_s} - 1}{10^{0.1\alpha_p} - 1} \tag{10.79}$$

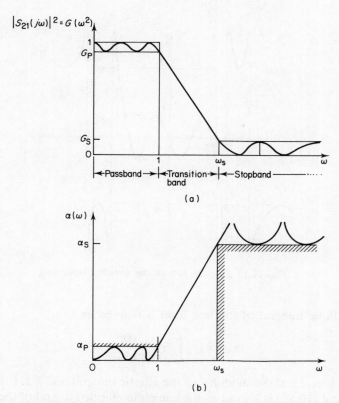

Fig. 10.10 Typical response of the optimum equiripple low-pass prototype low-pass filter. (a) Transducer power gain, (b) attenuation.

and

$$\varepsilon^2 = 10^{0.1\alpha_p} - 1 \qquad (10.80)$$

From the above requirements we must have

$$\frac{\mathrm{d}F_n}{\mathrm{d}\omega}\bigg|_{|F_n(\omega)|=1} = 0 \qquad \text{except at } |\omega| = 1 \qquad (10.81a)$$

$$\frac{\mathrm{d}F_n}{\mathrm{d}\omega}\bigg|_{|F_n(\omega)|=M^{-1}} = 0 \qquad \text{except at } |\omega| = \omega_s \qquad (10.81b)$$

Hence $F_n(\omega)$ satisfies the differential equation

$$\frac{\mathrm{d}F_n(\omega)}{\sqrt{\{1-F_n^2(\omega)\}\{M^2-F_n^2(\omega)\}}} = \frac{A \, \mathrm{d}\omega}{\sqrt{(1-\omega^2)(\omega_s^2-\omega^2)}} \qquad (10.82)$$

where A is a constant. The solution of the above equation can be obtained in terms of the elliptic integrals and Jacobian elliptic functions. These are now introduced very briefly.[8,61]

216

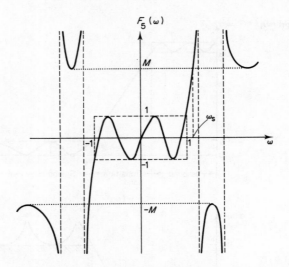

Fig. 10.11 Typical plot of the required behaviour
of $F_n(\omega)$, $n = 5$.

The elliptic integral of the first kind is defined as

$$u(\phi, k) = \int_0^\phi \frac{dx}{\sqrt{1 - k^2 \sin^2 x}} \qquad (10.83)$$

in which k is called the modulus of the elliptic integral and $k \le 1$. If $\phi = \pi/2$, the integral (10.83) is known as the complete elliptic integral of the first kind denoted by

$$K(k) = u\!\left(\frac{\pi}{2}, k\right) = \int_0^{\pi/2} \frac{dx}{\sqrt{1 - k^2 \sin^2 x}} \qquad (10.84)$$

If we let

$$k' = \sqrt{1 - k^2} \qquad (10.85a)$$

then

$$K'(k) = u\!\left(\frac{\pi}{2}, k'\right) = \int_0^{\pi/2} \frac{dx}{\sqrt{1 - k'^2 \sin^2 x}} \qquad (10.85b)$$

which is the complementary elliptic integral of the first kind. The elliptic integrals can be obtained from extensive tables in mathematical handbooks.[62] Alternatively, a simple computer program can be written to evaluate these integrals.[61,63]

Now the Jacobian elliptic functions are defined with reference to the notation in (10.83) to (10.85) as follows

$$\begin{aligned} \text{elliptic sine} \quad & sn(u, k) = \sin \phi \\ \text{elliptic cosine} \quad & cn(u, k) = \cos \phi \end{aligned} \qquad (10.86)$$

and the elliptic tangents, etc., can be defined in a similar manner. The values of these functions can also be obtained from tables[62] or by writing a simple computer program.[61,63]

Returning now to the solution of the differential equation (10.82), it can be shown[8,61] that $F_n(\omega)$ takes one of the following forms according to whether n is even or odd,

(i) *n even*

$$F_n(\omega) = C_e \prod_{r=1}^{n/2} \left\{ \frac{\omega^2 - \omega_r^2}{\omega^2 - (\omega_s^2/\omega_r^2)} \right\} \tag{10.87a}$$

where

$$\omega_r = \text{sn}\left(\frac{(2r-1)K(\omega_s^{-1})}{n}, \omega_s^{-1} \right) \tag{10.87b}$$

$$C_e = \prod_{r=1}^{n/2} \left\{ \frac{1 - (\omega_s^2/\omega_r^2)}{1 - \omega_r^2} \right\} \tag{10.87c}$$

(ii) *n odd*

$$F_n = C_0 \omega \prod_{r=1}^{(n-1)/2} \left\{ \frac{\omega^2 - \omega_r^2}{\omega^2 - (\omega_s^2/\omega_r^2)} \right\} \tag{10.88a}$$

where

$$\omega_r = \text{sn}\left(\frac{2rK(\omega_s^{-1})}{n}, \omega_s^{-1} \right) \tag{10.88b}$$

$$C_0 = \prod_{r=1}^{(n-1)/2} \left\{ \frac{1 - (\omega_s^2/\omega_r^2)}{1 - \omega_r^2} \right\} \tag{10.88c}$$

In the above expressions C_e and C_0 are constants, determined such that $F_n(1) = 1$.

In order to determine the degree of the elliptic filter required to meet a given set of specifications α_p, α_s, and ω_s the following formula may be used.

$$n \geq \frac{K(\omega_s^{-1})K'(M^{-1})}{K'(\omega_s^{-1})K(M^{-1})} \tag{10.89}$$

where M is obtained from α_p, α_s and (10.79). Naturally, the passband edge is at $\omega = 1$. Alternatively, the specifications may be given (or re-expressed) in terms of the minimum passband return loss α_{Rp}, minimum stopband attenuation α_s, and the ratio of stopband to passband edges $\gamma \equiv \omega_s$. Then a useful approximate design formula[8] is given by

$$n \geq \frac{\alpha_{Rp} + \alpha_s + 12}{13.65} \frac{K(\omega_s^{-1})}{K'(\omega_s^{-1})} \tag{10.90}$$

Now consider a comparison between the degrees of the maximally flat,

Chebyshev, and elliptic filters required to meet the same set of specifications:

> minimum passband return loss: 16.4 dB
> minimum stopband attenuation: 30 dB
> ratio of stopband to passband edge frequencies: 1.3

Substituting in (10.49), (10.77), and (10.90) and using the tables for elliptic integrals we obtain

> $n = 21$ for a maximally flat filter
> $n = 8$ for a Chebyshev filter
> $n = 5$ for an elliptic filter

which illustrates the superiority of the elliptic filter over the other types.

We now turn to the problem of the synthesis of the elliptic transfer function. We first note that when (10.87) or (10.88) are substituted in (10.78), the transfer function has all its transmission zeros on the $j\omega$-axis; these are all finite with an additional zero at ∞ for n odd. The realization of these finite $j\omega$-axis zeros of transmission requires, in general, the use of Brune sections of the type shown in Fig. 6.6. However, we have also seen in Section 7.5 that if the input impedance, formed from $|S_{21}(j\omega)|^2$ using the standard procedure in Section 6.3.2, satisfies the conditions of (Fujisawa's) Theorem 7.1 or Corollary 7.1 then the finite zeros of transmission may be realized without coupled coils. This results in the mid-shunt or mid-series forms of Fig. 7.5. Fujisawa's conditions were studied to investigate the realizability of the low-pass elliptic filter as a mid-shunt or mid-series ladder. The results may be summarized by stating that synthesis without the aid of coupled coils *cannot* be accomplished in the following cases[28]

(a) α_p is too small for a prescribed α_s; or
(b) α_s is too small for a prescribed α_p; or
(c) the transition band ($\omega_s - 1$) is made too broad for a prescribed α_p (or α_s).

Fortunately, it turns out that most practical specifications require transfer functions which are realizable using the mid-series or mid-shunt ladders discussed in Theorem 7.1 and Corollary 7.1 of Section 7.5. The required resistor-terminated ladder is then obtained by realizing the input impedance as explained in the proof of Theorem 7.1. There are, however, some special considerations associated with the even-degree case, but this matter will not be pursued any further. Due to the fact that elliptic filters provide the optimum amplitude response for a given degree, the normalized design has been made available in the form of an extensive set of tables.[64] These may be used directly to obtain the element values of the elliptic ladder filter for a given set of specifications.

10.2.5 Frequency transformations and impedance scaling

So far we have considered protoype low-pass filters where cutoff frequencies (i.e. passband edges) were normalized to unity ($\omega = 1$) and the source

Table 10.1 Frequency transformations and impedance scaling. In the last two columns ω_0 and β are given by (10.96); ω_s is the stopband edge in the low-pass prototype.

Normalized Low-Pass Prototype cutoff, $\omega=1$	Low-pass cutoff, $\omega=\omega_0$	High-pass cutoff, $\omega=\omega_0$	Band-pass passband: $\omega_1 \rightarrow \omega_2$	Band-stop stopband: $\omega_1 \rightarrow \omega_2$
L_r	$\dfrac{L_r}{\omega_0}$	$\dfrac{1}{L_r\omega_0}$	$\dfrac{\beta L_r}{\omega_0}$ $\dfrac{1}{\beta L_r\omega_0}$	$\dfrac{L_r\omega_s}{\beta\omega_0}$ $\dfrac{\beta}{L_r\omega_0\omega_s}$
C_r	$\dfrac{C_r}{\omega_0}$	$\dfrac{1}{C_r\omega_0}$	$\dfrac{1}{\beta C_r\omega_0}$ $\dfrac{\beta C_r}{\omega_0}$	$\dfrac{\beta}{C_r\omega_0\omega_s}$ $\dfrac{C_r\omega_s}{\beta\omega_0}$

For denormalization to arbitrary source resistor R. $L \rightarrow RL$, $C \rightarrow C/R$, $r_i \rightarrow Rr_i$

resistor was also taken as $1\,\Omega$. We now discuss the transformation to arbitrary cutoff as well as the high-pass, band-pass, and band-stop cases. Denormalization to an arbitrary source resistor is also explained. Table 10.1 gives a summary of these procedures which are derived below.

LOW-PASS TO LOW-PASS TRANSFORMATION

In order to have the same low-pass characteristic of the prototype but change the cutoff from $\omega=1$ to any arbitrary value $\omega=\omega_0$ we apply the transformation

$$p \rightarrow p/\omega_0 \tag{10.91}$$

in the transfer function. In terms of the element values in the realized prototype (e.g. Fig. 10.7) this amounts to scaling all frequency-dependent elements by $1/\omega_0$, i.e.

$$L_r \rightarrow L_r/\omega_0$$
$$C_r \rightarrow C_r/\omega_0 \tag{10.92}$$

LOW-PASS TO HIGH-PASS TRANSFORMATION

To obtain a high-pass filter with cutoff at ω_0 from a low-pass prototype we first let $p \rightarrow 1/p$ in the prototype transfer function. This transforms the low-pass response into a high-pass one with cutoff at $\omega=1$, and amounts to replacing every capacitor of value C by an inductor of value $1/C$ while every inductor of value L becomes a capacitor of value $1/L$. Next, for a cutoff at $\omega=\omega_0$ the same transformations in (10.91) to (10.92) are applied. Therefore, to obtain a high-pass filter with cutoff at ω_0 from a low-pass prototype network (e.g. Fig. 10.7) the elements are transformed according to

$$L_r \rightarrow C_r' = 1/\omega_0 L_r$$
$$C_r \rightarrow L_r' = 1/\omega_0 C_r \tag{10.93}$$

220

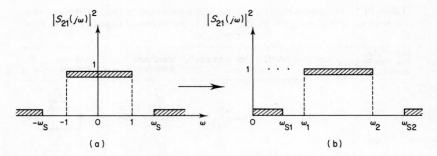

Fig. 10.12 Low-pass to band-pass transformation. (a) Low-pass prototype specifications, (b) band-pass specifications.

LOW-PASS TO BAND-PASS TRANSFORMATION

Consider the low-pass prototype specifications Fig. 10.12(a) with the negative frequency side also shown. We now seek a transformation of the prototype filter, into a band-pass one with the passband extending from ω_1 to ω_2 as shown in Fig. 10.12(b). This can be achieved via the transformation

$$p \rightarrow \beta\left(\frac{p}{\omega_0}+\frac{\omega_0}{p}\right) \tag{10.94}$$

which implies that, in the low-pass prototype network every inductor is transformed into a series resonant circuit, and every capacitor into a parallel resonant circuit. To evaluate β and ω_0 from ω_1 and ω_2 we write by reference to Fig. 10.12

$$-1=\beta\left(\frac{\omega_1}{\omega_0}-\frac{\omega_0}{\omega_1}\right)$$
$$1=\beta\left(\frac{\omega_2}{\omega_0}-\frac{\omega_0}{\omega_2}\right) \tag{10.95}$$

Simultaneous solution of the above equations gives

$$\omega_0=\sqrt{\omega_1\omega_2}, \qquad \beta=\frac{\omega_0}{\omega_2-\omega_1} \tag{10.96}$$

For example, the low-pass Chebyshev response function given by (10.50) is transformed into a band-pass function to give

$$|S_{21}(j\omega)|^2=\frac{1}{1+\varepsilon^2 T_n^2\left\{\beta\left(\frac{\omega}{\omega_0}-\frac{\omega_0}{\omega}\right)\right\}} \tag{10.97}$$

For design purposes, it is generally more convenient to transform the band-pass specifications into equivalent low-pass prototype specifications for the purpose of determining the required degree and element values. Then the transformations in Table 10.1 are used to obtain the band-pass filter.

LOW-PASS TO BAND-STOP TRANSFORMATION

In this case, we first transform the low-pass prototype into a high-pass prototype with stopband edge at $\omega = 1$. Next the same band-pass transformation (10.94) is applied to the high-pass prototype to obtain a band-stop filter with stopband extending from ω_1 to ω_2.

IMPEDANCE SCALING

For an arbitrary source resistor R, all impedance values are scaled by the same value R. Thus $L \to RL$, $C \to C/R$, and $r_\ell \to Rr_\ell$. Naturally if the prototype had equal terminations, then $r_\ell(=1\,\Omega) \to R$.

Example 10.1 Design a maximally-flat low-pass filter with the following specifications:

Passband: 0–10 kHz, attenuation ≤ 3 dB
Stopband edge: 40 kHz, attenuation ≥ 30 dB
Equal terminating resistors of 600 Ω

Solution The normalized value of stopband edge is $\omega_s = 40/10 = 4$, relative to the 3-dB point. Thus, the required degree of the filter is obtained from (10.47) as

$$n \geq \frac{30}{20 \log 4} \geq 2.49$$

$$n = 3$$

The element values of the prototype are obtained from (10.43) and this is shown in Fig. 10.13(a). Next this network is impedance-scaled by 600 Ω and the transformation (10.92) is used to locate the cutoff at $\omega_0 = 2\pi \times 10^4$. Thus

$$L_1 = L_3 = \frac{1 \times 600}{2\pi \times 10^4} = 9.55 \text{ mH}$$

$$C_2 = \frac{2}{600 \times 2\pi \times 10^4} = 0.053 \ \mu\text{F}$$

The final network is shown in Fig. 10.13(b).

Example 10.2 Design a Chebyshev low-pass filter with the following

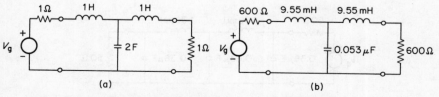

Fig. 10.13   Filter of Example 10.1. (a) Normalized prototype, (b) required filter.

222

specifications:

    Passband: 0–10 kHz, 0.0988 dB ripple
    Stopband edge: 20 kHz, attenuation ≥ 30 dB
    Equal terminating resistors of 50 Ω

Solution Normalizing the frequencies to the given cutoff of 10 kHz we have $\omega_s = 20/10 = 2$. Also from (10.74),

$$\varepsilon^2 = 10^{0.00988} - 1 = 0.023$$

or

$$\varepsilon = 0.152$$

Substituting in (10.76) for $\alpha_s = 30$, $\alpha_p = 0.0988$, and $\omega_s = 2$ we obtain $n = 5$. The auxiliary parameter in (10.64) is

$$\eta = \sinh\left\{\frac{1}{5}\sinh^{-1}\frac{1}{0.153}\right\} = 0.543$$

Therefore using (10.71) we obtain the element values of the prototype with reference to Fig. 10.7(b), $n = 5$. These are

$$C_1 = C_5 = 1.144$$
$$L_2 = L_4 = 1.372$$
$$C_3 = 1.972$$

Finally the above values are scaled in impedance by 50 Ω and the transformation (10.92) is used to give the required cutoff at 10 kHz. This yields

$$C_1' = C_5' = \frac{1.144}{50 \times 2\pi \times 10^4} = 0.36 \ \mu F$$

$$L_2' = L_4' = \frac{1.372 \times 50}{2\pi \times 10^4} = 1.09 \ mH$$

$$C_3' = \frac{1.972}{50 \times 2\pi \times 10^4} = 0.63 \ \mu F$$

The required filter is shown in Fig. 10.14.

Example 10.3 Design a band-pass Chebyshev filter with the specifications illustrated by the tolerance scheme of Fig. 10.15 with 50 Ω equal terminations

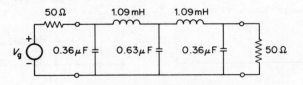

Fig. 10.14 Filter of Example 10.2.

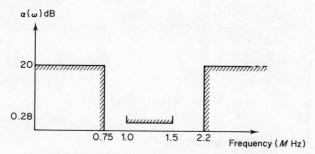

Fig. 10.15   Specifications on the filter of Example 10.3.

Solution   The transformation of (10.94) and Fig. 10.12 produces a response with geometric symmetry around band-centre ω_0. $\alpha(\omega)$ has the same value at every pair of frequencies f_{s1}, f_{s2} related by $f_{s1} f_{s2} = f_0^2$. The specifications of Fig. 10.15 do not possess such symmetry: $1.0 \times 1.5 \neq 0.75 \times 2.2$. Therefore, with $f_0^2 = 1.0 \times 1.5 = 1.5$, the filter has to be designed according to the more severe of the two requirements: 20 dB at 0.75 MHz or 20 dB at 2.2 MHz, and the other one will be over-satisfied. If we require $\alpha \geq 20$ for $f \geq 2.2$, we also obtain $\alpha \geq 20$ for $f \leq (1.5/2.2) \leq 0.68$ and we have failed to satisfy the lower stopband requirement at 0.75. On the other hand, requiring $\alpha \geq 20$ for $f \leq 0.75$ also gives $\alpha \geq 20$ for $f \geq (1.5/0.75) \geq 2$. Thus, the upper stopband requirement at 2.2 is over-satisfied, therefore, we use the 20 dB requirement at 0.75 to determine the prototoype. From (10.96),

$$\omega_0 = \sqrt{\omega_1 \omega_2} = 2\pi \times 1.225 \times 10^6$$

$$(\omega_2 - \omega_1) = \pi \times 10^6$$

$$\beta = \frac{\omega_0}{\omega_2 - \omega_1} = 2.45$$

The frequency 0.75 MHz corresponds to $-\omega_s$ in Fig. 10.12(a) of the low-pass prototype, and is obtained using (10.94) as

$$-\omega_s = 2.45 \left(\frac{0.75}{1.225} - \frac{1.225}{0.75} \right) = -2.5$$

also
$$\varepsilon^2 = 10^{0.028} - 1 = 0.0666$$

Using (10.8) and (10.50), the above value corresponds to a passband return loss of

$$\alpha_{Rp} = 10 \log \left(\frac{1 + \varepsilon^2}{\varepsilon^2} \right) = 12 \text{ dB}$$

Therefore by (10.77) the degree of the prototype is $n = 3$. The element values, with reference to Fig. 10.7(b) with $n = 3$, are $C_1 = 1.345$, $L_2 = 1.141$, $C_2 = 1.345$. The transformation in (10.94), or Table 10.1, is used with scaling by 50 Ω to obtain the required 6th order band-pass filter of Fig. 10.16.

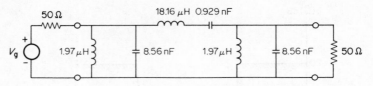

Fig. 10.16   Filter of Example 10.3.

10.3   PHASE APPROXIMATION

10.3.1   Statement of the problem

In the previous section we gave solutions to the amplitude approximation problem. We now turn to the problem of designing networks with prescribed phase (or group-delay) characteristics in the passband. We shall be only concerned with approximations to the ideal linear phase characteristic shown in Fig. 10.3. We also restrict our attention to functions of the form

$$S_{21}(p) = \frac{1}{Q(p)} \tag{10.98}$$

so that it is realizable as a simple low-pass ladder of the general form in Fig. 10.7. Write

$$Q(p) = M(p) + N(p) \tag{10.99}$$

where M is even and N is odd. Then the phase function of $S_{21}(p)$ is

$$\psi(p) = -\tanh^{-1} \frac{N(p)}{M(p)}, \qquad p \to j\omega \tag{10.100a}$$

so that,

$$\psi(\omega) = -\tan^{-1} \frac{N(j\omega)}{jM(j\omega)} \tag{10.100b}$$

The phase $\psi(\omega)$ is required to be linear in the passband, alternatively the group-delay should approximate to a constant in the passband, since

$$T_g(p) = -\frac{d\psi(p)}{dp} \tag{10.101a}$$

and

$$T_g(\omega) = -\frac{d\psi(\omega)}{d\omega} \tag{10.101b}$$

10.3.2   Maximally flat group-delay response

There are several methods[8,65] for the derivation of the nth degree polynomial $Q(p)$ in (10.99) which results in a group-delay function $T_g(\omega)$ having the maximum number of zero derivatives at $\omega = 0$. A possible method is to write

$$\tanh \psi(p) = \frac{Q_n - Q_{n*}}{Q_n + Q_{n*}}$$

$$= \frac{N(p)}{M(p)} \tag{10.102}$$

and if the right-hand side of (10.102) were identical with (sinh p/cosh p) then the delay is constant at all frequencies. However, since N and M are real polynomials we must find a method for approximating (sinh p/cosh p) $\equiv$ tanh p by a rational function. An obvious way is to write

$$\sinh p = p + \frac{p^3}{3!} + \frac{p^5}{5!} + \ldots$$
$$\cosh p = 1 + \frac{p^2}{2!} + \frac{p^4}{4!} + \ldots \tag{10.103}$$

and a continued fraction expansion of (sinh p/cosh p) gives

$$\tanh p = \cfrac{1}{\cfrac{1}{p} + \cfrac{1}{\cfrac{3}{p} + \cfrac{1}{\cfrac{5}{p} + \cfrac{\ddots}{\cfrac{1}{\cfrac{(2n-1)}{p} + \ddots}}}}}$$

$$\tag{10.104}$$

Now, a maximally flat group-delay around $p = 0$ results if N/M in (10.102) is identified with the nth approximant in the continued fraction expansion of tanh p as given by (10.104). This means that we truncate the expansion at the nth step, then the resulting terms are remultiplied to form a rational function approximation to tanh p. This is equated to N/M in (10.102) and the required polynomial $Q_n(p)$ in (10.99) is defined. This procedure gives in closed form,

$$Q_n(p) = \sum_{k=0}^{n} b_k p^k \tag{10.105}$$

where

$$b_k = \frac{\binom{n}{k}}{(2n)!} 2^k (2n - k)! \tag{10.106}$$

Alternatively $Q_n(p)$ may be generated by means of the recurrence formula

$$Q_{n+1}(p) = Q_n(p) + \frac{p^2}{(4n^2 - 1)} Q_{n-1}(p) \qquad (10.107)$$

with

$$Q_0 = 1, \quad Q_1 = 1 + p$$

In fact, the polynomial $Q(p)$ is related to the well known Bessel polynomial $B(p)$ by

$$Q(p) = p^n B_n(1/p) \qquad (10.108)$$

and the realized network is sometimes called a *Bessel filter*. The maximally flat character of the group-delay of $S_{21}(p)$ can be demonstrated using (10.101) and the properties of $Q(p)$. Thus

$$T_g(p) = \mathrm{Ev}\left\{\frac{Q'}{Q}\right\}$$

$$= 1 - \left\{\frac{(-1)^n}{(2n-1)} \prod_1^{n-1} \frac{1}{(4r^2 - 1)}\right\} \frac{p^{2n}}{Q_n Q_{n*}} \qquad (10.109)$$

from which $T_g(p)$ has $(2n - 1)$ zero derivatives at $p = 0$.

Since $Q_n(p)$ is obtained according to (10.104), it is strictly Hurwitz, and $|S_{21}(j\omega)| \leq 1$ for all ω. The transfer function has all its zeros (of transmission) at $p = \infty$, therefore it is realizable in the simple ladder form of Fig. 10.7 using the technique of Section 7.2.

10.3.3 Equidistant linear phase approximation

By contrast with the amplitude case, the equiripple solution to the group-delay approximation problem, is non-optimum.[8] The numerical solution to this problem, however, provides an improvement over the maximally flat case, but has been superseded by a more satisfactory analytical solution which satisfies a strong optimality criterion. Rhodes[8] has shown that the *equiripple* solution to the problem of *minimizing the maximum deviation* from *phase linearity* over a finite band *is optimum*. This is obtained by deriving the polynomial which *interpolates the ideal phase angle* $\psi(\omega) = \omega$ over a finite band as shown in Fig. 10.17. This satisfies

$$A_n(j\omega \mid \alpha) = A(\omega) \exp\{j\psi(\omega)\} \qquad (10.110)$$

where

$$\{\omega - \psi(\omega)\}_{\omega = r\alpha} = 0 \qquad r = 0, 1, 2, \ldots, n \qquad (10.111)$$

Therefore, the phase is exactly linear at equidistant frequency increments. A_n is called the *equidistant linear phase polynomial*, and can be generated by means of the recurrence formula[8,66]

$$A_{n+1}(p \mid \alpha) = A_n(p \mid \alpha) + \frac{\tan^2 \alpha}{\alpha^2} \cdot \frac{p^2 + (\alpha n)^2}{(4n^2 - 1)} A_{n-1}(p \mid \alpha) \qquad (10.112)$$

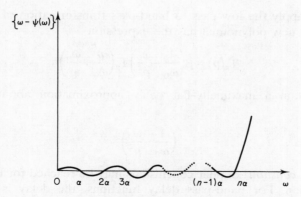

Fig. 10.17 Phase error in the equidistant linear phase approximation.

with

$$A_0(p \mid \alpha) = 1, \quad A_1(p \mid \alpha) = 1 + \left(\frac{\tan \alpha}{\alpha}\right) p$$

The necessary and sufficient condition for $Q_n(p \mid \alpha)$ to be strictly Hurwitz is

$$\alpha < \frac{\pi}{2} \qquad (10.113)$$

Clearly, if $\alpha \to 0$ in (10.112) the maximally flat solution is recovered as given in (10.107). Finally, the transfer function formed as

$$S_{21}(p) = \frac{1}{A(p \mid \alpha)} \qquad (10.114)$$

is realizable as a low-pass ladder of the general form shown in Fig. 10.7 using the standard synthesis steps (6.22) to (6.27) of Section 6.3.2.

10.3.4 Are frequency transformations possible?

A low-pass to low-pass transformation of delay-oriented designs is possible using (10.91) and (10.92), to adjust the passband edge to any arbitrary value. Impedance scaling by the source resistance can also be applied. However, the transformations to *high-pass and band-pass, etc. cannot be used* since the delay characteristics are *distorted* by such transformations. To illustrate this point consider the group delay of the maximally-flat delay polynomial given by (10.107). From (10.109) the power series expansion of the delay function is of the form

$$T_g(p) = 1 - a_1 p^{2n} - a_2 p^{2n+2} \ldots \qquad (10.115)$$

or the phase is of the form

$$\psi(p) = -p + a_1' p^{2n+1} + a_2' p^{2n+3} \qquad (10.116)$$

Now if we apply the low-pass to band-pass transformation of (10.94), the delay of the new polynomial has the expression

$$\hat{T}_{\mathrm{g}}(p) = \beta\left(\frac{1}{\omega_0} - \frac{\omega_0}{p^2}\right)T_{\mathrm{g}}\left(\frac{p}{\omega_0} + \frac{\omega_0}{p}\right) \tag{10.117}$$

and we obtain a maximally-flat delay approximation about ω_0 to the function

$$\left(\frac{1}{\omega_0} - \frac{\omega_0}{p^2}\right) \tag{10.118}$$

and *not to a constant*! Similar conclusions may be reached for the high-pass transformation. For band-pass delay functions, the delay is required to possess the power series expansion

$$T_{\mathrm{g}}(p) = 1 + a_1(\omega^2 - \omega_0^2)^n + a_2(\omega^2 - \omega_0^2)^{n+1} + \dots \tag{10.119}$$

However, there is no known analytical solution to this problem, although it may be solved numerically as an eigenvalue problem.[67] More recently, further solutions have been proposed and the problem is still the subject of active research.[68]

10.4 SIMULTANEOUS AMPLITUDE AND PHASE APPROXIMATION

10.4.1 Statement of the problem

We now turn to the problem of designing low-pass filters which exhibit good amplitude and phase characteristics *simultaneously*. A possible approach is to design the filter on amplitude basis (e.g. Chebyshev or elliptic), subsequently we phase-equalize the resulting network by cascading it with an all-pass section. The phase response of the all-pass phase-equalizer is determined such that it compensates for the deviation, from phase linearity, of the amplitude-oriented filter. The general properties of all-pass two-ports were discussed in Section 6.7. Alternatively linear phase networks can be amplitude-equalized. However, in both approaches the solutions are far from optimum, i.e. the degree of the network is higher than one would anticipate by examining the available degrees of freedom. For this reason, only solutions to the problem of simultaneous amplitude and phase approximations are pointed out here. These require the constraints on both responses to be imposed together at the outset for the derivation of the required transfer function.

We have seen in Section 10.1.2 that for a minimum-phase transfer function, good amplitude and good phase characteristics are not compatible. In fact, Carlin and Wu[69] gave a very elegant investigation of the trade-off between amplitude selectivity and phase linearity in minimum-phase prototype functions. The trade-off was given in terms of bandwidth restriction on linear phase functions with amplitude selectivity. It was shown that a minimum-phase lossless two-port with monotonic stopband loss and whose amplitude selectivity is *prescribed*, has its linear phase region *limited* to a certain bandwidth *independent* of the precise realization of the network or its complexity. A technique was also given for obtaining the optimum compromise.

From the above discussion, it follows that if highly selective, linear phase characteristics are required, the minimum-phase constraint (10.15) must be removed, and general non-minimum-phase transfer functions are used. Unfortunately, to date the optimum solution to the combined amplitude and phase problem has not been found. However, there exists a large number of useful solutions differing in the method of formulation and in the relative emphasis laid on either the amplitude or the phase.[70-73] In fact, most of these solutions can be derived as special degenerate cases of the corresponding solutions in the distributed and digital domains to be discussed in the later chapters. Therefore, only some solutions are mentioned here, without derivation, which employ transfer functions with even numerators, and hence realizable directly in reciprocal form. For more general cases the reader may consult the references.[70-73] Regarding the synthesis techniques of the transfer functions which are presented below, the procedure of cascade synthesis in Chapter 6 can always be applied.

10.4.2 Optimum maximally flat passband amplitude and linear phase response

Consider a transfer function with an even numerator for a reciprocal realization (see Section 5.5, Theorem 5.3),

$$S_{21}(p) = \frac{E_{2m}(p)}{D_n(p)} \tag{10.120}$$

which is required to possess the following properties:

(a) Maximally-flat amplitude around $p = 0$, i.e. $|S_{21}(j\omega)|^2$ has $(2n-1)$ zero derivatives at $\omega = 0$.
(b) $(m+1)$ zero derivatives of the group-delay at $\omega = 0$.
(c) A single transmission zero at $p = \infty$.
(d) $S_{21}(0) = 1$.

Note that by allowing $S_{21}(p)$ to possess a numerator which is a *non-trivial even polynomial*, by contrast with the cases discussed hitherto, we are able to impose the conditions on both the amplitude and delay simultaneously. The available degrees of freedom are greater than in the all-pole functions discussed before.

Now the function which possesses the properties (a) to (d) above was obtained by Rhodes.[8] For n odd we have $(n = 2m + 1)$,

$$S_{21}(p) = \frac{\text{Ev}\{Q_{m*}(2Q_{m+1} - Q_m)\}}{Q_m\{2Q_{m+1} - Q_m\}} \tag{10.121}$$

where Q is the maximally flat delay polynomial defined by (10.105) to (10.107).

The even-degree function with the same properties except that it possesses two zeros of transmission at $p = \infty$ is given by

$$S_{21}(p) = \frac{E_{2m-2}(p)}{D_{2m}(p)} \tag{10.122}$$

where

$$E_{2m-2}(p) = Q_m Q_{m*} + \frac{p^2}{(2m-1)^2} Q_{m-1} Q_{m-1*} \tag{10.123}$$

$$D_{2m}(p) = Q_m^2 + \frac{p^2}{(2m-1)^2} Q_{m-1}^2 \tag{10.124}$$

and Q_m is the same polynomial defined by (10.105) to (10.107).

Now the direct synthesis of the present transfer functions will require zero-sections (see Section 6.5), and this is now illustrated by an example.

Example 10.4 Realize the 4th degree transfer function whose general expression is given by (10.122) to (10.124).

Solution Here $m = 2$ and from (10.107),

$$Q_1 = 1 + p, \quad Q_2 = 1 + p + \tfrac{1}{3}p^2$$

Substituting for Q_1 and Q_2 in (10.122) to (10.124) with $m = 2$ we obtain

$$S_{21}(p) = \frac{9 - 2p^2}{9 + 18p + 16p^2 + 8p^3 + 2p^4}$$

which possesses two zeros of transmission at $p = \infty$, realizable as a series inductor and a shunt capacitor. It also has a pair of finite real-axis zeros at $p = \pm 3/\sqrt{2}$ which require a C-section. Following steps (1) to (5) in (6.23) to (6.26) of Section 6.3.2 we have

$$S_{11}S_{11*} = 1 - S_{21}S_{21*}$$

$$= \frac{4p^8}{81 - 36p^2 + 4p^4 + 4p^8}$$

Thus

$$S_{11} = \frac{2p^4}{9 + 18p + 16p^2 + 8p^3 + 2p^4}$$

and the input impedance is given by

$$Z(p) = \frac{1+S_{11}}{1-S_{11}}$$

$$= \frac{4p^4+8p^3+16p^2+18p+9}{8p^3+16p^2+18p+9}$$

The Brune preamble (Section 6.4) is performed to extract all $j\omega$-axis poles from Z or Y. In the present case Z has a pole at $p=\infty$ which is extracted by an inductor of value 0.5 H. The remainder is

$$Z_1 = Z-0.5p$$

$$= \frac{7p^2+13.5p+9}{8p^3+16p^2+18p+9}$$

and $Y_1 = Z_1^{-1}$ has a pole at $p=\infty$ which is extracted as a shunt capacitor of value $\frac{8}{7}$F. The remainder is

$$Y_2 = Y_1 - \frac{8}{7}p$$

$$= \frac{4p^2+54p+63}{49p^2+94.5p+63}$$

which is minimum-reactance and minimum susceptance. We may now apply cascade synthesis (Section 6.5) to realize $Z_2(p) = Y_2(p)^{-1}$. The numerator of Ev Z_2 is $(9-2p^2)^2$ as expected from $S_{21}(p)$. Thus, we require a C-section to realize the transmission zeros at $p=\sigma_0 = \pm 3/\sqrt{2}$. Using (6.42), (6.45) evaluate $Z_2(3/\sqrt{2}) = 2.477$ and $Z_2'(3/\sqrt{2}) = 0.648$. Finally the element values of the C-section of Fig. 6.6 are obtained from the formulae in (6.61). The complete filter is shown in Fig. 10.18.

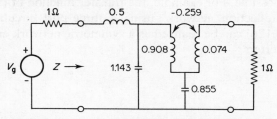

Fig. 10.18   Filter of Example 10.4, element values in Henrys and Farads.

10.4.3   Finite-band approximation to constant amplitude and linear phase

Use can be made of the equidistant linear phase polynomial $A(p \mid \alpha)$ defined by (10.112) to obtain transfer functions satisfying the following constraints

$$S_{21}(\infty) = 0 \tag{10.125a}$$

$$|S_{21}(j\omega)| \leq 1 \tag{10.125b}$$

$$S_{21}(jr\beta) = e^{jr\alpha} \qquad r = 1, 2, \ldots, n \tag{10.125c}$$

where

$$\beta = \tan \alpha \qquad (10.125d)$$

i.e. the amplitude is unity at a number of equidistant frequencies, and the phase interpolates to the linear characteristic at regular frequency intervals. The odd degree transfer function with an even numerator is of the form[8]

$$S_{21} = \frac{E_{2m}(p)}{G_m(p)H_{m+1}(p)} \qquad (10.126)$$

where

$$G_m(p) = A_m(p \mid \alpha) \qquad (10.127)$$

$$H_{m+1}(p) = (1+k)A_{m+1}(p \mid \alpha) - kA_m(p \mid \alpha) \qquad (10.128)$$

$$E_{2m}(p) = \text{Ev}\{G_m*H_{m+1}\} \qquad (10.129)$$

and k is a constant, which may be chosen to provide an *extra* phase or amplitude constraint, e.g. a zero of transmission at $p = \infty$. Letting

$$\gamma_r = \left\{\frac{A_r(p \mid \alpha)}{A_{r-1}(p \mid \alpha)}\right\}_{p=j\omega_{r-1}} \qquad r = 1, 2, \ldots, m \qquad (10.130)$$

then the conditions for a bounded real S_{21} in (10.126) are

$$\gamma_r \geq 0, \quad k \geq -1 \qquad r = 1, 2, \ldots, m \qquad (10.131)$$

The synthesis may be performed using the general techniques of Chapter 6 as in the previous subsection.

Regarding the general approach to the synthesis of the present functions as well as those of Section 10.4.2, it-is noteworthy that alternative techniques are also possible. For example, the transfer function of (10.121) can be realized as a reflection network[8] using a three-port circulator. Also the function in (10.126) can be realized as a symmetric network employing ideal impedance inverters.[8]

PROBLEMS

10.1 Design a low-pass maximally flat filter with the following specifications:

> Passband: 0–1 MHz, attenuation ≤ 3 dB
> Stopband edge: 1.5 MHz, attenuation ≥ 40 dB
> Equal terminating resistors of 50 Ω.

10.2 Design a low-pass Chebyshev filter with the following specifications:

> Passband: 0–0.5 MHz, 0.2 dB ripple
> Stopband edge: 1 MHz, attenuation ≥ 50 dB
> Equal terminating resistors of 75 Ω.

10.3 Design a band-pass maximally flat filter with the following specifications:

Passband: 10 kHz–15 kHz, attenuation ≤ 3 dB
Stopband edges at 9 kHz and 17 kHz, minimum attenuation of 30 dB in both stopbands
Equal terminating resistors of 600 Ω.

10.4 Design a band-pass Chebyshev filter with the following specifications:

Passband: 10 kHz–15 kHz, 0.28 dB ripple
Stopband edges at 8.5 kHz and 17 kHz, minimum attenuation of 40 dB in both stopbands
Equal terminating resistors of 50 Ω.

10.5 Calculate the degrees of elliptic filters which would meet the same specifications of Problems 10.1 to 10.4.

10.6 Design a maximally flat delay (Bessel) low-pass prototype filter with a maximum variation of group-delay in the passband of 1 per cent. The passband edge is defined at $\omega = 1$. Also obtain the ladder filter which meets the same specifications but employing the equidistant linear phase polynomial. Compare the degrees of the two networks. In each case, equal terminating resistors of 1 Ω are assumed.

Chapter 11

Approximation Methods for Commensurate Distributed Filters

11.1 INTRODUCTION

As shown in Chapter 8, commensurate distributed networks are describable by rational or quasi-rational functions of the Richards' variable

$$\lambda = \tanh \tau p$$
$$= \Sigma + j\Omega \tag{11.1}$$

in which τ is the one-way delay of the UE. Thus, the approximation problem in the present case consists in the derivation of a rational (or quasi-rational) transfer function of the doubly terminated lossless two-port, such that it satisfies certain specifications on its amplitude and/or phase characteristics. The transfer function of the required lossless two-port shown in Fig. 11.1 may be written as

$$S_{21} = \frac{N_m(\lambda)}{D_n(\lambda)} \qquad m \leq n \tag{11.2}$$

where $D_n(\lambda)$ is a strictly Hurwitz polynomial, and $N(\lambda)$ is a real polynomial apart from a possible factor $(1-\lambda^2)^{1/2}$. Clearly $S_{21}(\lambda)$ must also satisfy

$$|S_{21}(j\Omega)| \leq 1 \qquad -\infty \leq \Omega \leq \infty \tag{11.3}$$

Moreover, for a reciprocal two-port $N(\lambda)$ must be even or odd, again apart from a possible factor $(1-\lambda^2)^{1/2}$.

11.2 AMPLITUDE APPROXIMATION

11.2.1 General considerations

The transducer power gain of a finite lossless commensurate distributed network may be written as

$$|S_{21}(j\Omega)|^2 = \frac{\displaystyle\sum_{r=0}^{m} a_r \Omega^{2r}}{\displaystyle\sum_{r=0}^{n} b_r \Omega^{2r}} \tag{11.4}$$

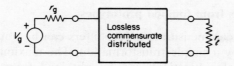

Fig. 11.1 The commensurate distributed filter.

where

$$\Omega = \tan \tau \omega \tag{11.5}$$

and the response is, clearly, periodic with ω. The ideal low-pass amplitude characteristic is shown in Fig. 11.2, which is often known as *quasi-low-pass* due to the periodic appearance of several passbands.

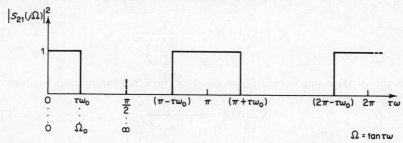

Fig. 11.2 Ideal quasi-low-pass amplitude response of commensurate distributed filters.

It is important to note that the transformation

$$\lambda \to 1/\lambda \tag{11.6}$$

in a low-pass transfer function provides a *band-pass* function with the passband centred at $\tau \omega = \pi/2$, with arithmetic symmetry around band-centre. The band-pass function has the same properties around $\tau \omega = \pi/2$ as those of the low-pass one around $\tau \omega = 0$. This is illustrated in Fig. 11.3, which may also be regarded as a *quasi-high-pass* response.

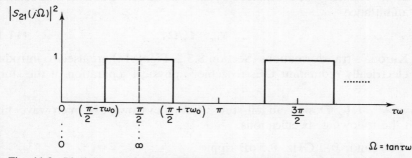

Fig. 11.3 Ideal band-pass (quasi-high-pass) amplitude response of commensurate distributed filters.

11.2.2 Synthesis from lumped prototypes

All-stub lossless commensurate ladder filters can be obtained from lumped lossless ladders by a simple transformation. The maximally flat, Chebyshev and elliptic transfer functions can be obtained from the corresponding lumped prototype functions by the transformation

$$p \rightarrow \lambda/\Omega_0 \tag{11.7}$$

where

$$\Omega_0 = \tan \tau\omega_0 \tag{11.8}$$

This transforms the point $\omega = 1$ in the lumped domain to $\Omega = \Omega_0 = \tan \tau\omega_0$ in the distributed domain.

Thus, for the distributed counterpart of the low-pass maximally-flat filter we have

$$|S_{21}(j\Omega)|^2 = \frac{1}{1 + \left(\dfrac{\Omega}{\Omega_0}\right)^{2n}} \tag{11.9}$$

with the 3-dB point occurring at ω_0. For the Chebyshev case we have

$$|S_{21}(j\Omega)|^2 = \frac{1}{1 + \varepsilon^2 T_n^2\left(\dfrac{\Omega}{\Omega_0}\right)} \tag{11.10}$$

The expressions for the elliptic transfer functions can be obtained in a similar manner from (10.87) to (10.88) and the transformation (11.7).

Consequently the all-stub ladder filter with cutoff ω_0 can be directly obtained from the lumped ladder prototype filter. This is accomplished as follows:

(a) Every inductor L_r is replaced by a short-circuited stub of characteristic impedance

$$Z_{0r} = L_r/\Omega_0 \tag{11.11}$$

(b) Every capacitor C_r is replaced by an open-circuited stub of characteristic admittance

$$Y_{0r} = C_r/\Omega_0 \tag{11.12}$$

(c) Kuroda's transformations (Section 8.5.2, Fig. 8.14) are used to introduce electrically redundant UEs to achieve physical separation of the stubs.

Example 11.1 Design an all-stub Chebyshev low-pass microwave filter with the following specifications:

Passband: 0–1 GHz, 0.5 dB ripple
Stopband: 3 GHz–5 GHz, attenuation ≥ 30 dB
Equal terminating resistors of 50 Ω.

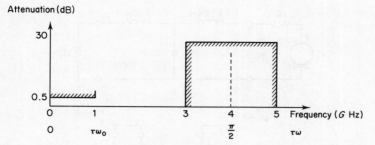

Fig. 11.4 Specifications on filter of Example 11.1.

Use Kuroda's transformations to introduce UEs between the stubs. Figure 11.4 illustrates the specifications as a tolerance scheme.

Solution We first determine the degree of the lumped prototype counterpart, required to meet the given specifications. In the passband, $0 \to \tau\omega_0$ we require

$$10 \log (1 + \varepsilon^2) \leq 0.5 \text{ dB}$$

i.e. $\varepsilon^2 = 0.122$, which corresponds to a return loss of

$$\alpha_{\mathrm{Rp}} = 10 \log \frac{1}{|S_{11R}|^2} = 10 \log \frac{1 + \varepsilon^2}{\varepsilon^2}$$

$$= 9.6 \text{ dB}$$

The one-way delay τ of the UE is obtained by noting that $\tau\omega = \pi/2$ corresponds to the stopband centre: 4 GHz. Thus

$$2\pi \times 4 \times 10^9 \tau = \frac{\pi}{2}$$

or

$$\tau = 6.25 \times 10^{-11} \text{ s}$$

The corresponding commensurate length of the UE is given by $l = c \cdot \tau$, where c is the velocity of propagation of electromagnetic waves in the medium. Taking this to be the speed of light, we obtain $l = 1.875$ cm. Next, from the stopband requirement, we require $\alpha_s \geq 30$ dB at 3 GHz. The ratio of stopband to passband frequencies in the Ω-domain is

$$\frac{\Omega_s}{\Omega_0} = \frac{\tan \tau\omega_s}{\tan \tau\omega_0} = 5.829$$

Substituting in (10.77) with $\omega \to \Omega/\Omega_0$ we obtain

$$n \geq \frac{30 + 9.6 + 6}{20 \log \{5.829 + \sqrt{(5.829)^2 - 1}\}} \geq 2.14$$

Therefore $n = 3$. The element values of the third order Chebyshev lumped

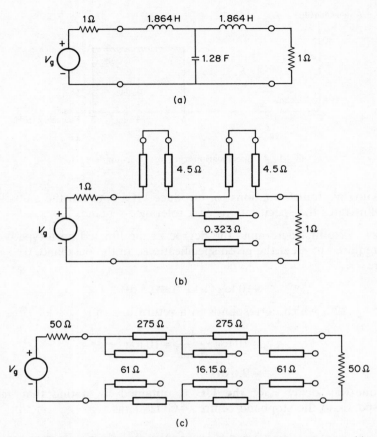

Fig. 11.5 Procedure for obtaining the required filter of Example 11.1. (a) Lumped prototype, (b) all-stub counterpart, (c) final required filter after applying Kuroda's transformations and scaling by 50 Ω.

prototype with $\varepsilon^2 = 0.122$ are obtained from (10.64) and (10.71). This is shown in Fig. 11.5(a). The element values in the all-stub ladder are obtained from (11.11) and (11.12) with $\Omega_0 = \tan \tau \omega_0 = 0.414$. This gives the network of Fig. 11.5(b). After impedance-scaling by 50 Ω and application of Kuroda's transformations, the final network of Fig. 11.5(c) results.

It must be noted that the UEs which are used to separate the stubs, by means of Kuroda's transformations, are redundant in an electrical sense, though necessary in a physical sense. This immediately poses the question as to whether it is possible to design filters containing stubs and UEs but in which the UEs contribute to the response as well as achieving physical separation of the stubs. We shall see that the solution to this problem does exist. Furthermore, a very important class of filters employs cascades of UEs *alone* without stubs, which can be realized very conveniently in practi᠁ This class is considered next.

11.2.3 Cascaded UE filters with maximally flat and equiripple passbands

The transfer function of a doubly terminated cascade of n UEs satisfies Corollary 8.1, expressions (8.34) and (8.35). Thus, it is of the form

$$S_{21}(\lambda) = \frac{(1-\lambda^2)^{n/2}}{D_n(\lambda)} \qquad (11.13)$$

where $D_n(\lambda)$ is strictly Hurwitz and $|S_{21}(j\Omega)|^2 \le 1$ for all Ω. Thus

$$|S_{21}(j\Omega)|^2 = \frac{(1+\Omega^2)^n}{|D_n(j\Omega)|^2} \qquad (11.14)$$

and using

$$\sin^2 \tau\omega = \frac{\Omega^2}{1+\Omega^2} \qquad (11.15)$$

$$\cos^2 \tau\omega = \frac{1}{1+\Omega^2} \qquad (11.16)$$

then (11.14) may be put in the form

$$|S_{21}|^2 = \frac{1}{1+P_n(\sin^2 \tau\omega)} \qquad (11.17)$$

Now, for a maximally flat response around $\tau\omega = 0$, and one zero derivative at $\tau\omega = \pi/2$ (i.e. $\Omega = \infty$) we have

$$|S_{21}|^2 = \frac{1}{1+\left(\dfrac{\sin \tau\omega}{\sin \tau\omega_0}\right)^{2n}} \qquad (11.18)$$

and the 3-dB point occurs at ω_0.

For an optimum equiripple response in the passband up to ω_0, and one zero-derivative at $\tau\omega = \pi/2$ ($\Omega = \infty$) we have

$$|S_{21}|^2 = \frac{1}{1+\varepsilon^2 T_n^2\left(\dfrac{\sin \tau\omega}{\sin \tau\omega_0}\right)} \qquad (11.19)$$

where

$$T_n\left(\frac{\sin \tau\omega}{\sin \tau\omega_0}\right) = \cos\left\{n \cos^{-1}\left(\frac{\sin \tau\omega}{\sin \tau\omega_0}\right)\right\}$$

$$= \cosh\left\{n \cosh^{-1}\left(\frac{\sin \tau\omega}{\sin \tau\omega_0}\right)\right\} \qquad (11.20)$$

which is the Chebyshev polynomial of the first kind in $(\sin \tau\omega/\sin \tau\omega_0)$. A typical response of the function in (11.19) is shown in Fig. 11.6.

The synthesis of the transfer functions in (11.18) or (11.19) can, in general

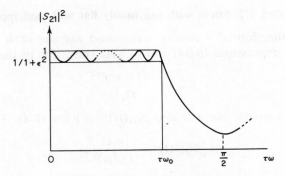

Fig. 11.6 Equiripple-passband response of a cascaded U.Es
filter.

be performed by the standard technique. Thus $|S_{11}|^2$ is formed, then let
$j\Omega \rightarrow \lambda$ and $S_{11}S_{11*}$ is factored to obtain a bounded real $S_{11}(\lambda)$. Finally the
input impedance $Z(\lambda)$ of the resistor-terminated cascade is formed using
(8.28) and realized using either repeated application of Richards' theorem
(see Example 8.1 in Section 8.5.1) or preferably the explicit formulae given
in expressions (8.37) to (8.43). It is to be noted that tables for the element
values in these UE cascades are also available.[74]

Example 11.2 Design a Chebyshev low-pass microwave filter as a cascade
of UEs with the following specifications

 (a) Passband: 0–1 GHz, 0.4 dB ripple
 (b) Stopband centred at 3 GHz with a minimum attenuation of 40 dB at
 this point.

Solution Passband ripple of 0.4 dB corresponds to $\varepsilon^2 = 0.1$. We now obtain
the design with 1 Ω equal terminations, which may be later impedance-
scaled by the actual value. From (b) the centre of the stopband at 3 GHz
corresponds to $\tau\omega = \pi/2$. Thus $2\pi \times 3 \times 10^9 \tau = \pi/2$, i.e. $\tau = 12 \times 10^{-9}$ s. The
required length of the commensurate lines is $l = c\tau = 2.5$ cm. From the
passband requirement $\tau\omega_0 = \pi/6$; thus

$$\sin \tau\omega_0 = \sin \pi/6 = 0.5$$

Therefore using (11.19) and the requirement (b),

$$10 \log \left\{ 1 + 0.1 T_n^2 \left(\frac{\sin \pi/2}{\sin \pi/6} \right) \right\} \geq 40$$

or

$$10 \log \{ 1 + 0.1 T_n^2(2) \} \geq 40$$

A useful property of the Chebyshev polynomial $T_n(x)$ is that for $x \gg 1$,
$T_n(x) \approx 2^{n-1} x^n$. Using this, we obtain $n \geq 4.66$, i.e. $n = 5$. The exact attenua-

tion at $\tau\omega = \pi/2$ can be checked using

$$T_5(x) = 16x^5 - 20x^3 + 5x, \quad x = 2$$

which gives 41.2 dB.

Having obtained the required degree of the filter we can proceed with the synthesis. Writing $|S_{21}|^2$ explicitly

$$|S_{21}|^2 = \frac{1}{1 + 0.1\{16(2\sin\tau\omega)^5 - 20(2\sin\tau\omega)^3 + 5(2\sin\tau\omega)\}^2}$$

we evaluate

$$|S_{11}|^2 = 1 - |S_{21}|^2$$

then use (11.15) to obtain $|S_{11}(j\Omega)|^2$. Next, let $j\Omega \to \lambda$ to give $S_{11}S_{11*}$ which is factored and a strictly Hurwitz denominator is selected. For the minimum-phase $S_{11}(\lambda)$ we have

$$S_{11}(\lambda) = \frac{114.5\lambda^5 + 44.27\lambda^3 + 3.16\lambda}{114.5\lambda^5 + 83.21\lambda^4 + 74.48\lambda^3 + 28.89\lambda^2 + 8.53\lambda + 1}$$

Finally the input impedance is obtained as

$$Z(\lambda) = \frac{1 + S_{11}(\lambda)}{1 - S_{11}(\lambda)}$$

which may be realized using repeated application of Richards' theorem or expressions (8.37) to (8.43). This gives the filter of Fig. 11.7, which may be impedance scaled by an arbitrary resistor value $r_g = r_\ell$

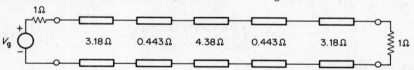

Fig. 11.7   Filter of Example 11.3.

11.2.4   Cascaded UEs and stubs

For optimum use of stubs and UEs, all elements must contribute to the response of the network. For a low-pass response with n UEs and q stubs (series short-circuited or shunt open-circuited) we have from (8.54),

$$|S_{21}(j\Omega)|^2 = \frac{(1 + \Omega^2)^n}{P_{n+q}(\Omega^2)} \tag{11.21}$$

Clearly q is the number of transmission zeros at $\tau\omega = \pi/2$ ($\Omega = \infty$). Equation (11.21) may be put in the form

$$|S_{21}|^2 = \frac{1}{1 + F_n^2\left(\dfrac{\sin\tau\omega}{\sin\tau\omega_0}\right)} \tag{11.22}$$

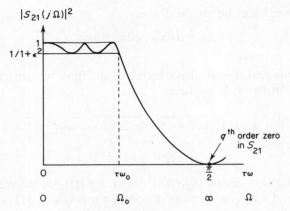

Fig. 11.8 Typical response of a filter described by
(11.22) and (11.23).

For an optimum equiripple passband up to ω_0 and q zero-derivatives at $\tau\omega = \pi/2$ $(\Omega = \infty)$ we have

$$F_n = \varepsilon \cosh \left\{ n \cosh^{-1}\left(\frac{\sin \tau\omega}{\sin \tau\omega_0}\right) + q \cosh^{-1}\left(\frac{\tan \tau\omega}{\tan \tau\omega_0}\right) \right\} \tag{11.23}$$

Figure 11.8 shows a typical response of the function defined by (11.22) and (11.23). For a maximally flat response around $\tau\omega = 0$ and q zero-derivatives at $\tau\omega = \pi/2$ $(\Omega = \infty)$,

$$F_n = \left(\frac{\sin \tau\omega}{\sin \tau\omega_0}\right)^n \left(\frac{\tan \tau\omega}{\tan \tau\omega_0}\right)^q \tag{11.24}$$

with the 3-dB point at ω_0.

The synthesis of the above functions is obtained by the usual method of forming the input impedance, then extracting the UEs and the stubs in any convenient order. Naturally, a UE is extracted by application of (Richards') Theorem 8.2, and a stub is extracted by removing a pole at $\lambda = \infty$ from the impedance or admittance.

The general form of the band-pass (or quasi-high-pass) function corresponding to (11.21) is obtained by letting $\Omega \to 1/\Omega$. This gives

$$|S_{21}|^2 = \frac{\Omega^{2q}(1+\Omega^2)^n}{P_{n+q}(\Omega^2)}$$

$$= \frac{1}{1+F_n^2} \tag{11.25}$$

For the equiripple case,

$$F_n = \varepsilon \cosh \left\{ n \cosh^{-1}\left(\frac{\cos \tau\omega}{\cos \tau\omega_0}\right) + q \cosh^{-1}\left(\frac{\cot \tau\omega}{\cot \tau\omega_0}\right) \right\} \tag{11.26}$$

whereas for the maximally flat case,

$$F_n = \left(\frac{\cos \tau\omega}{\cos \tau\omega_0}\right)^n \left(\frac{\cot \tau\omega}{\cot \tau\omega_0}\right)^q \qquad (11.27)$$

In both cases (11.26) and (11.27), the passband is centred at $\tau\omega = \pi/2$ and extends from

$$\tau\omega_0 \to (\pi - \tau\omega_0) \qquad (11.28)$$

Note that the q transmission zeros at $\lambda = 0$ are realized by shunt short-circuited or series open-circuited stubs.

Finally, the transfer functions which are realizable as interdigital networks can be obtained by taking $q = 1$ in (11.25) to (11.27). The synthesis of such functions was discussed in Section 8.6.2.

Example 11.3 Realize the band-pass function defined by (11.25) and (11.26) with $\varepsilon^2 = 0.02$, $n = 2$, $q = 1$, and $\tau\omega_0 = \pi/3$. The corresponding response is shown in Fig. 11.9(a).

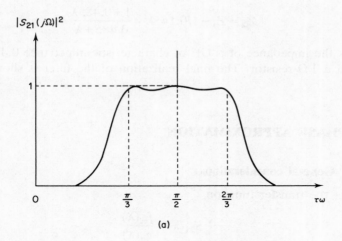

(a)

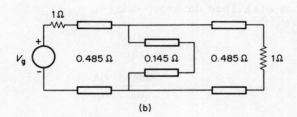

(b)

Fig. 11.9   Filter of Example 11.3. (a) Response, (b) realized filter.

Solution The required transducer power gain is given by

$$|S_{21}|^2 = \frac{1}{1 + 0.02\cos^2\{2\cosh^{-1}(2\cos\tau\omega) + \cosh^{-1}(1.732\cot\tau\omega)\}}$$

Forming the input reflection coefficient $S_{11}(\lambda)$ from $|S_{21}|^2$ in the usual manner and selecting a minimum-phase $S_{11}(\lambda)$ we calculate the input impedance of the 1 Ω-terminated two-port. This gives

$$Z(\lambda) = \frac{4.344\lambda + 2.592\lambda^2 + \lambda^3}{6.896 + 4.344\lambda + 4.124\lambda^2 + \lambda^3}$$

We already know from the form of $|S_{21}|^2$ that the filter contains two UEs and a single stub. First extract a UE of characteristic impedance $Z_{01} = Z(1) = 0.485\ \Omega$. The remainder is calculated using (8.30) as

$$Z_1(\lambda) = \frac{\lambda + 0.485\lambda^2}{6.895 + 4.344\lambda + 2.062\lambda^2}$$

Thus $Y_1 = Z_1^{-1}$ has a pole at $\lambda = 0$ which is extracted as a shunt short-circuited stub of characteristic impedance $Z_{02} = 1/6.895 = 0.145\ \Omega$. The remainder is given by

$$Z_2 = Z_1 - 1/0.145\lambda = \frac{1 + 0.485\lambda}{0.485 + \lambda}$$

which is the impedance of a UE of characteristic impedance 0.485, terminated in a 1 Ω-resistor. The final realization of the filter is shown in Fig. 11.9(b).

11.3 PHASE APPROXIMATION

11.3.1 General considerations

Consider the transfer function

$$S_{21}(\lambda) = \frac{f_m(\lambda)}{g_n(\lambda)} \tag{11.29}$$

whose phase is $\psi(\lambda)$. Then the group-delay is

$$\begin{aligned}
T_g(\lambda) &= -\frac{d\psi(\lambda)}{dp} \\
&= -\frac{d\psi(\lambda)}{d\lambda} \cdot \frac{d\lambda}{dp} \\
&= -\tau(1 - \lambda^2)\frac{d\psi}{d\lambda} \tag{11.30}
\end{aligned}$$

The factor $(1 - \lambda^2)$ in the above expression makes lumped prototypes of no use in distributed (and digital) delay approximations. Furthermore, frequency scaling by a factor $(\lambda \rightarrow k\lambda)$ is not possible here since $(1 - \lambda^2)$ does not scale. This necessitates the incorporation of a scaling parameter in the expressions for transfer functions at the outset.

Using similar analysis to that in Section 10.1.2 we obtain

$$T_g(\lambda) = \text{Ev} \left\{ (1 - \lambda^2) \left(\frac{g'}{g} - \frac{f'}{f} \right) \right\} \tag{11.31}$$

Now, a band-pass (or quasi-high-pass) transfer-function can be obtained from the low-pass $S_{21}(\lambda)$ by letting

$$\lambda \rightarrow 1/\lambda \tag{11.32}$$

The above transformation results in a delay function as in (11.31) but with $\lambda \rightarrow 1/\lambda$; thus no dispersion occurs by contrast with the lumped case. Therefore the delay properties around $\tau\omega = 0$ are transformed to $\tau\omega = \pi/2$ $(\Omega = \infty)$ unaffected.

Consider a transfer function of the *specific form*

$$S_{21}(\lambda) = \frac{E_m(\lambda)}{(1 + \lambda)^n} \tag{11.33}$$

where $E_m(\lambda)$ is an *even* polynomial and, consequently does not contribute to the delay variation. Substitution in (11.31) gives for the specific form (11.33),

$$T_g(\lambda) \equiv n \tag{11.34}$$

i.e. the *delay is a constant (or the phase is exactly linear) at all frequencies*.

As in the case of lumped networks the phase approximation problem reduces to that of determining a polynomial

$$Q(\lambda) = M(\lambda) + N(\lambda) \tag{11.35}$$

where $M(\lambda)$ is even and $N(\lambda)$ is odd, such that

$$\psi(\lambda) = \tanh^{-1} \frac{N(\lambda)}{M(\lambda)} \tag{11.36}$$

approximates to the desired characteristics when used in (11.30).

11.3.2 Maximally flat group-delay response

The polynomial $Q_n(\lambda)$ which gives rise to a group-delay function with $(2n - 1)$ zero-derivatives at $\lambda = 0$ can be generated by the recurrence relationship[75]

$$Q_{n+1}(\lambda) = Q_n(\lambda) + \frac{\alpha^2 - n^2}{4n^2 - 1} \lambda^2 Q_{n-1} \tag{11.37}$$

with

$$Q_0 = 1, \quad Q_1 = 1 + \alpha\lambda$$

where α is a *bandwidth scaling parameter*, which may be interpreted as the delay at the origin ($\lambda = 0$). The polynomial $Q(\lambda)$ is obtained by identifying $N(\lambda)/M(\lambda)$ in (11.25) with the nth approximant in a continued fraction expansion of

$$\tanh(\alpha\tan^{-1}\lambda) = \frac{\sinh(\alpha\tanh^{-1}\lambda)}{\cosh(\alpha\tanh^{-1}\lambda)} \qquad (11.38)$$

The condition for $Q_n(\lambda)$ to be strictly Hurwitz is

$$\alpha \geq n - 1 \qquad (11.39)$$

The simplest way to form a transfer function with maximally flat group-delay around the origin, is to write

$$S_{21}(\lambda) = \frac{1}{Q(\lambda)} \qquad (11.40)$$

and the maximally flat character of the group-delay can be proved by deriving an expression analogous to (10.109) in the lumped case.

It is also possible to show that $Q_n(\lambda)$ is related to the symmetrical Jacobi polynomial[8] $P_n(\lambda)$ by

$$P_n(\lambda) = \lambda^n Q_n(1/\lambda) \qquad (11.41)$$

which results in a maximally flat delay around $\lambda = \infty$ ($\tau\omega = \pi/2$) and hence gives the solution to the corresponding band-pass (or quasi-high-pass) case. Note that *this result has no counterpart in the lumped domain*.

11.3.3 Equidistant linear-phase approximation

Rhodes[8] has derived a polynomial $A_n(\lambda \mid \phi_0)$ which satisfies

$$\arg A_n(\pm j\tan r\phi_0 \mid \phi_0) = \pm\beta r\phi_0 \qquad r = 0, 1, 2, \ldots, n \qquad (11.42)$$

with the associated phase error function

$$\varepsilon(\tau\omega) = \beta\omega - \psi(\tan\tau\omega) \qquad (11.43)$$

having the typical behaviour shown in Fig. 11.10. The phase is, therefore, exactly linear at a set of equidistant points $0, \phi_0, 2\phi_0, \ldots, n\phi_0$ in the passband. The required polynomial can be obtained by the recurrence formula

$$A_{n+1}(\lambda \mid \phi_0) = A_n(\lambda \mid \phi_0) + \gamma_n(\lambda^2 + \tan^2 n\phi_0)A_{n-1}(\lambda \mid \phi_0) \qquad (11.44)$$

with

$$A_0(\lambda \mid \phi_0) = 1, \quad A_1(\lambda \mid \phi_0) = 1 + \frac{\tan\beta\phi_0}{\tan\phi_0}\lambda$$

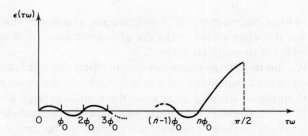

Fig. 11.10 Phase error of the distributed equidistant linear
phase polynomial

and

$$\gamma_n = \frac{\cos{(n-1)\phi_0}\cos{(n+1)\phi_0}\sin{(\beta+n)\phi_0}\sin{(\beta-n)\phi_0}\cos^2{n\phi_0}}{\sin{(2n-1)\phi_0}\sin{(2n+1)\phi_0}\cos^2{\beta\phi_0}}$$

(11.45)

If we form a transfer function as

$$S_{21} = \frac{1}{A_n(\lambda \mid \phi_0)}$$

(11.46)

then S_{21} is bounded real if

$$\beta \geq n-1, \quad n\phi_0 \leq \frac{\pi}{2}, \quad (\beta+n-1)\phi_0 \leq \pi$$

(11.47)

11.4 SIMULTANEOUS AMPLITUDE AND PHASE APPROXIMATION

As in the case of lumped networks, it is possible to design the filter on amplitude basis, then use all-pass sections as phase-equalizers. Alternatively, linear phase filters may be amplitude-equalized. However, this solution to the combined amplitude and phase approximation problem is far from optimum.

For minimum-phase transfer functions, considerations similar to those given in Section 10.1.2 for the lumped case lead to the conclusion that amplitude selectivity and phase linearity are staunch opponents. In fact, this problem has been examined[76] for the quasi-minimum-phase case of transfer functions realizable as cascades of UEs. It has been shown that once the amplitude selectivity is specified, there is an upper bound on the fraction of passband over which linear phase response can be maintained. For a given selectivity, this upper bound cannot be exceeded no matter how many UEs are employed. A design technique to approximately realize this upper bound was given.[76] Other techniques for simultaneous amplitude and phase approximation for this class of networks are also available.[70] We also note the work by Rhodes on interdigital filters.[8]

If we remove the minimum-phase constraint, then the available degrees of freedom are increased considerably. Consequently, in the most general case of non-minimum-phase transfer functions, the available degrees of freedom can, in principle, be divided arbitrarily between the amplitude and phase responses. Examples of these solutions were given in the lumped domain in Section 10.4 which may be generalized to commensurate distributed networks.[70,77,78] However, in their most general form non-minimum phase transfer functions of $\lambda = \tanh \tau p$ require fairly complex realizations. But the analogy between digital and commensurate distributed transfer functions, makes these solutions of direct practical use as digital transfer functions. Examples of these solutions will be given in the next chapter. Furthermore, as pointed out earlier, it is possible to obtain distributed transfer functions which have exact linear phase at all frequencies. These must be of the form given by (11.33) and will be treated in the next chapter since they can be conveniently realized as FIR digital filters.

PROBLEMS

11.1 Design a microwave all-stub ladder Chebyshev filter with the following specifications

> Passband: 0–1 GHz, 0.5 dB ripple
> Stopband: 1.5 GHz–3.5 GHz, attenuation ≥ 30 dB
> Equal terminating resistors of 50 Ω.

Use Kuroda's transformations to introduce UE between the stubs.

11.2 Design a Chebyshev microwave filter as a cascade of UEs which meets the following specifications

> Passband: 0–1.5 GHz, 0.2 dB ripple
> Stopband: 2 GHz–3 GHz, attenuation ≥ 40 dB
> Equal terminating resistors of 75 Ω.

11.3 Design an interdigital (band-pass) filter with optimum equiripple passband response to meet the following specifications:

> Passband: 0.5 GHz–0.9 GHz, 0.5 dB ripple
> Stopband edges at 0.2 GHz and 1.2 GHz attenuation ≥ 30 dB in both stopbands.

11.4 Consider the transducer power gain of a microwave band-pass filter,

$$|S_{21}|^2 = \frac{1}{1 + 0.1 \cos^2 \tau\omega \cot^2 \tau\omega}$$

Find the realization and sketch the amplitude response for $\tau = 2ns$.

Chapter 12
Approximation Methods for Digital Filters

12.1 GENERAL TECHNIQUE

We have seen in Chapter 9 that the transfer function of a digital filter may be expressed as

$$H(z) = \frac{\sum\limits_{r=0}^{m} a_r z^{-r}}{1 + \sum\limits_{r=1}^{n} b_r z^{-r}} \tag{12.1}$$

where

$$z^{-1} = e^{-Tp}, \ T = 1/f_N = 2\pi/\omega_N \tag{12.2}$$

and f_N is the sampling (Nyquist) frequency. Alternatively we may employ the variable

$$\lambda = \tanh \frac{T}{2} p \tag{12.3}$$

and for $p \to j\omega$, $\lambda \to j\Omega$ where

$$\Omega = \tan \frac{T}{2} \omega = \tan \pi \frac{\omega}{\omega_N} \tag{12.4}$$

Clearly λ and z are related by

$$\lambda = \frac{1 - z^{-1}}{1 + z^{-1}} \tag{12.5}$$

$$z^{-1} = \frac{1 - \lambda}{1 + \lambda} \tag{12.6}$$

Thus, the transfer function of the digital filter may be expressed as

$$H(\lambda) = \frac{N(\lambda)}{D(\lambda)} \tag{12.7}$$

and, again, all responses of digital filters are periodic with ω as explained in

249

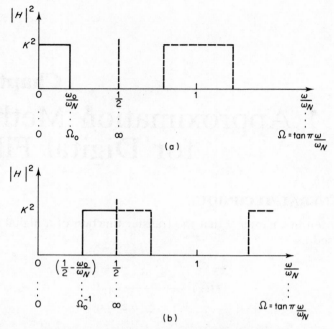

Fig. 12.1 Ideal amplitude response of a digital filter. (a) Low-pass, (b) high-pass.

Section 9.4. The ideal low-pass amplitude characteristic is shown in Fig. 12.1(a). Having studied the approximation problem in commensurate distributed networks, and established the analogy with the transfer functions of digital filters in Chapter 9, we may state the following important conclusion.

Strong Theorem 12.1 *Apart from the non-reciprocal nature of the digital filter, the solutions to the low-pass approximation problem in the digital domain are identical to those in the commensurate distributed domain with $\tau \to T/2$, i.e. Richards' variable in the digital case is given by (12.3). More specifically, any commensurate distributed low-pass transfer function $S_{21}(\lambda)$ can be regarded as a digital transfer function $H(\lambda)$ with the same amplitude and phase characteristics. Furthermore, the transformation $\lambda \to 1/\lambda$ in a low-pass function produces a high-pass one with the end of the passband at $\omega_N/2$ (Fig. 12.1(b))*

Consequently, all the low-pass transfer functions obtained in Chpater 11 can be realized in digital form if we take τ to represent $T/2$, i.e. half the sampling period in the digital domain. Naturally, the transformation (12.5) can be used to re-express $H(\lambda)$ as $H(z)$, if the realization requires it to be in the form (12.1). The synthesis techniques of digital transfer functions were discussed in Chapter 9, and the reader is referred to the relevant sections of that chapter. We now put these results together for direct use as digital transfer functions. We also augment the results by stating some solutions to the combined amplitude and phase approximation problem.

12.2 IIR FILTERS

12.2.1 Amplitude approximation

Letting $p \to \lambda/\Omega_0$ in the maximally flat, Chebyshev or elliptic lumped filter transfer functions we were able to obtain the corresponding distributed ones. Thus we can derive the digital counterparts by the transformation

$$p \to \frac{\lambda}{\Omega_0} \to \frac{1}{\Omega_0}\frac{1-z^{-1}}{1+z^{-1}} \tag{12.8}$$

where

$$\Omega_0 = \tan \pi \frac{\omega_0}{\omega_N} \tag{12.9}$$

which transforms the point $\omega = 1$ in the lumped prototype transfer function into ω_0 in the digital domain. In (12.9) $\omega_N = 2\pi f_N$, which is the radian sampling frequency.

MAXIMALLY FLAT RESPONSE IN BOTH PASSBAND AND STOPBAND

Letting $j\Omega \to \lambda$ in (11.9) we obtain $S_{21}S_{21*}$ which is factored, and a strictly Hurwitz denominator in λ is chosen. This gives

$$S_{21}(\lambda) \to H(\lambda) = \frac{K}{\displaystyle\prod_{r=1}^{n}\left\{\frac{\lambda}{\Omega_0} - je^{j\theta}r\right\}} \tag{12.10}$$

where

$$\theta_r = \frac{(2r-1)}{2n}\pi \qquad r = 1, 2, \ldots, n \tag{12.11}$$

and K is chosen such that the dC-gain is any prescribed value; $K = 1$ for $S_{21}(0) = 1$. ω_0 is the point where the gain falls by 3 dB. In the z-domain we use (12.5) to give

$$H(z) = \frac{K(1+z^{-1})^n}{\displaystyle\prod_{r=1}^{n}\left\{\left(\frac{1}{\Omega_0} - je^{j\theta}r\right) - z^{-1}\left(\frac{1}{\Omega_0} + je^{j\theta}r\right)\right\}} \tag{12.12}$$

CHEBYSHEV RESPONSE

Letting $j\Omega \to \lambda$ in (11.10) we obtain $S_{21}S_{21*}$ which is factored for a strictly

Hurwitz denominator in λ. This gives

$$S_{21}(\lambda) \to H(\lambda) = \frac{K}{\prod\limits_{r=1}^{n} \left\{ \dfrac{\lambda}{\Omega_0} + (\eta \sin \theta_r + j\sqrt{1+\eta^2} \cos \theta_r) \right\}} \tag{12.13}$$

where θ_r is given by (12.10) and

$$\eta = \sinh \left(\frac{1}{n} \sinh^{-1} \frac{1}{\varepsilon} \right) \tag{12.14}$$

Again, (12.5) may be used in (12.13) to obtain for the Chebyshev filter

$$H(z) = \frac{K(1+z^{-1})^n}{\prod\limits_{r=1}^{n} \left\{ \left(\dfrac{1}{\Omega_0} + jy_r \right) - z^{-1} \left(\dfrac{1}{\Omega_0} - jy_r \right) \right\}} \tag{12.15}$$

where

$$y_r = \cos \{ \sin^{-1} j\eta + \theta_r \} \tag{12.16}$$

Expressions (12.11) and (12.14) can be used directly to realize the functions in cascade form. Naturally, two factors combine to form a second order function, with a possible first order section for n odd.

Explicit expressions may also be derived for the elliptic filter, but these are best obtained from the extensive set of tables for lumped filters and using the same transformation (12.8).

DIGITAL EQUIVALENTS OF UE CASCADES WITH OR WITHOUT STUBS

Expressions (11.18) and (11.19) give amplitude low-pass functions realizable as cascades of UEs, while (11.22) to (11.24) employ both stubs and UEs in cascade. All these functions can be realized in digital form by first factoring the appropriate expression in the λ-domain, choosing a strictly Hurwitz denominator, then applying the transformation (12.5).

HIGH-PASS AND BAND-PASS FILTERS

In the treatment of commensurate distributed filters in Chapter 11, we were able to obtain quasi-high-pass and band-pass transfer functions from the quasi-low-pass ones by the transformation

$$\lambda \to 1/\lambda \tag{12.17}$$

This can only be used in the digital domain to obtain high-pass transfer functions with the stipulation that the passband-end is at $\omega_N/2$, i.e. the sampling frequency is twice the highest frequency in the useful pass-band.

On the other hand, the transformation (12.17) cannot be used to produce a band-pass response from a low-pass one in the digital domain. This is because the transformation places the 'hypothetical' band-centre at $\omega_N/2$ which is the end of the useful band beyond which aliasing occurs (see Fig.

12.1(b)) However, it is possible to use a low-pass to band-pass transformation similar to that in (10.94) with p replaced by λ. Thus, in the transfer functions of lumped low-pass prototypes we let

$$p \to \lambda \to \frac{\bar{\Omega}}{\Omega_2 - \Omega_1} \left(\frac{\lambda}{\bar{\Omega}} + \frac{\bar{\Omega}}{\lambda} \right) \qquad (12.18)$$

where

$$\bar{\Omega} = \tan \pi \frac{\bar{\omega}}{\omega_N} \qquad (12.19a)$$

$$\Omega_{1,2} = \tan \pi \frac{\omega_{1,2}}{\omega_N} \qquad (12.19b)$$

$$\bar{\Omega} = \sqrt{\Omega_1 \Omega_2} \qquad (12.19c)$$

The band-centre $\bar{\omega}$ is arbitrary, but the end of the upper stopband is at $\omega_N/2$. The passband extends from ω_1 to ω_2. Finally (12.5) may be used to transform to the z-domain.

12.2.2 Phase approximation

The polynomial $Q(\lambda)$ which gives a maximally flat group-delay around $\lambda = 0$ is given by (11.37). This can be used directly to form a digital transfer function with the same properties,

$$H(\lambda) = \frac{K}{Q(\lambda)} \qquad (12.20)$$

K is a constant and the transformation (12.5) is used to express H as a function of z. It is to be noted that the parameter α in the expressions for $Q(\lambda)$ in (11.37) is used for bandwidth scaling to an arbitrary point. Similarly, the distributed equidistant linear phase polynomial defined by (11.44) may be used directly to form a digital transfer function

$$H(\lambda) = \frac{K}{A(\lambda \mid \phi_0)} \qquad (12.21)$$

K is a constant. It must be emphasized that *digital delay functions* can only be obtained from *commensurate distributed transfer functions*, never from lumped prototype functions.

Again, the transformation $\lambda \to 1/\lambda$ can be used in (12.20) or (12.21) to obtain a high-pass delay response, with the end of the passband at $\omega_N/2$.

12.2.3 A little reminder of the synthesis

The synthesis techniques of the transfer functions discussed so far, were given in detail in Chapter 9. These are of three main types.

(a) Direct realizations as shown in Fig. 9.5.

(b) Cascade realizations as shown in Fig. 9.8. Alternatively the parallel form of realization of Fig. 9.9 may be used.

(c) Wave digital realizations for the purpose of imitating the low sensitivity properties of the classical passive structures. These are obtained by first realizing a *distributed reference network*, then the various elements are replaced by their wave digital equivalents with the connections effected by adaptors. This technique was discussed in detail in Section 9.7, with illustrative examples. We also note that the transfer functions in (12.10), (12.13), (12.20), and (12.21) are realizable, in the distributed domain, as all-stub ladders. Therefore the wave digital equivalents are obtained by the technique of Section 9.7.4. The wave digital realization of UE cascades described by (11.18) and (11.19) was also discussed in Section 9.7.4.

Example 12.1 Design a Chebyshev low-pass digital filter with the following specifications:

> Passband: 0–1 kHz, 0.5 dB ripple
> Stopband: 1.5 kHz–3 kHz, attenuation ≥ 30 dB.

Solution First, the degree of the required filter must be determined. Note that the amplitude function is of the form

$$|H(j\Omega)|^2 = \frac{1}{1 + \varepsilon^2 T_n^2\left(\dfrac{\Omega}{\Omega_0}\right)}$$

where Ω is given by (12.4) and Ω_0 by (12.9) with ω_0 as the passband edge. From the given specifications $\alpha_p = 0.5$ corresponds to

$$\varepsilon^2 = 10^{0.1\alpha_p} - 1 = 0.122$$

$$\varepsilon = 0.35$$

which, in turn represents a return loss of

$$\alpha_{Rp} = 10 \log\left(\frac{1 + \varepsilon^2}{\varepsilon^2}\right) = 9.6 \text{ dB}$$

The highest frequency in the stopband is 3 kHz; therefore the sampling frequency must be at least twice this value,

$$f_N \geq 6 \text{ kHz}$$

Taking $f_N = 6$ kHz, we have from (12.4) and (12.9)

$$\Omega_0 = \tan \pi \frac{\omega_0}{\omega_N} = \tan \pi \frac{f_0}{f_N} = \tan \frac{\pi}{6}$$

$$\Omega_s = \tan \pi \frac{\omega_s}{\omega_N} = \tan \frac{\pi}{4}$$

Therefore, the ratio of stopband to passband frequencies in the Ω-domain is

$$\frac{\Omega_s}{\Omega_0} = 1.732$$

We can now substitute in (10.77) with $\omega \to \Omega/\Omega_0$, and $\alpha_{Rp} = 9.6$, $\alpha_s = 30$. This gives

$$n \geq \frac{\alpha_{Rp} + \alpha_s + 6}{20 \log \left\{ \frac{\Omega_s}{\Omega_0} + \sqrt{\left(\frac{\Omega_s}{\Omega_0}\right)^2 - 1} \right\}}$$

$$\geq 4.58$$

i.e. $n = 5$. Therefore, the required transfer function is that of (12.13) or (12.15) with $n = 5$ and the auxiliary parameter η is given by

$$\eta = \sinh \left(\frac{1}{5} \sinh^{-1} \frac{1}{0.35} \right)$$

$$= 0.362$$

Having obtained the explicit expression of the function in terms of z, it can be realized in direct, cascade, or parallel form. Alternatively an all-stub reference ladder can be obtained, then a wave digital realization is constructed.

12.2.4 Simultaneous amplitude and phase approximation

The general form of a selective linear phase digital transfer function is identical to the corresponding distributed case. However, a distinct advantage of a digital realization is the increased degrees of freedom available in the transfer function so that we may write

$$H(\lambda) = \frac{N(\lambda)}{D(\lambda)} \tag{12.22}$$

where $N(\lambda)$ is an arbitrary general numerator, not restricted to be even or odd as for a reciprocal distributed transfer function. Thus, the non-reciprocal nature of the digital filter may be exploited to increase the available parameters (the coefficients of the numerator). Therefore, we have more flexibility for the shaping of both amplitude and phase responses simultaneously, without having to worry about complicated structures for the realization, or using an involved realizability theory.

There are many sophisticated analytical (and numerical) techniques for simultaneous amplitude and phase approximation.[8,70,73,77,78] They differ in the method of formulation as well as the relative degree of emphasis on either response. Only two representative classes will be given here for the low-pass case. The reader is urged to consult the references for other techniques of this important aspect of approximation theory.

Consider a transfer function, expressed as in (12.22) which is required to satisfy

$$H(j \tan r\phi_0) = e^{-j2\beta r\phi_0} \qquad r = 0, 1, \ldots, (n-1) \tag{12.23a}$$

$$H(\infty) = 0 \tag{12.23b}$$

i.e. a transmission zero at $\lambda = \infty$ ($\omega = \omega_N/2$) as well as possessing an equidistant interpolation to unity amplitude and linear phase responses. $H(\lambda)$ can be expressed in terms of the distributed equidistant linear phase polynomial $A_n(\lambda \mid \phi_0)$ given by (11.44) to (11.45). This gives

$$H(\lambda) = \frac{A_{n*} + k\lambda A_{n-1*}}{A_n + k\lambda A_{n-1}} \tag{12.24}$$

and k is chosen to produce a zero of transmission at $\lambda = \infty$ ($\omega = \omega_N/2$). Thus

$$k = \frac{A_n}{\lambda A_{n-1}} \bigg|_{\lambda \to \infty} \tag{12.25}$$

Next, a transfer function with simultaneous maximally flat amplitude and delay responses around $\lambda = 0$ can be obtained from (11.24) by letting $\phi_0 \to 0$. This gives the maximally flat delay polynomial $Q(\lambda)$ defined by (11.37). The transfer function is, therefore, given by

$$H(\lambda) = \frac{Q_{n*} + k\lambda Q_{n-1*}}{Q_n + kQ_{n-1}} \tag{12.26}$$

and k is chosen to produce a zero of transmission at $\lambda = \infty$ ($\omega = \omega_N/2$). Thus

$$k = \frac{Q_n}{\lambda Q_{n-1}} \bigg|_{\lambda \to \infty} \tag{12.27}$$

Finally, we note that both A_n and Q_n contain parameters for bandwidth scaling.

12.3 FIR FILTERS WITH EXACT LINEAR PHASE

It was shown in Section 9.5 that a digital filter of the FIR type has a transfer function of the form

$$H(z) = \sum_{r=0}^{n} a_r z^{-r} \tag{12.28}$$

i.e. it is a polynomial in z^{-1}, therefore it is unconditionally stable. Its direct non-recursive realization was discussed in Section 9.6.2 and is shown in Fig. 9.10. In terms of the λ-variable, (12.28) takes the general form

$$H(\lambda) = \frac{f_m(\lambda)}{(1+\lambda)^n} \tag{12.29}$$

and as shown in Section 11.3.1, if f_m is an even polynomial the filter has a constant group-delay (i.e. exact linear phase) at all frequencies. In (12.28) this implies that $H(z)$ is a symmetric polynomial in z^{-1}.

Once the exact linear phase property is maintained by choosing f_m as an even polynomial, our efforts can be entirely concentrated on the amplitude approximation problem. There is no known analytical solution to the optimum equiripple approximation problem; however, it has been solved numerically.[79,80] The monotonic response with the optimum number of constraints, divided arbitrarily between the passband and stopband, can be easily stated.[8,81] In this case the available number of zero derivatives of $|H(j\Omega)|^2$ must be capable of being divided arbitrarily between $\Omega = 0$ and $\Omega = \infty$ (i.e. $\omega = 0$ and $\omega = \omega_N/2$). Thus, e.g. for $n = 2k$, i.e. even,

$$H(\lambda) = \frac{\sum_{r=0}^{m} (-1)^r \binom{k}{r} \lambda^{2r}}{(1+\lambda)^{2k}} \tag{12.30}$$

so that in HH_* the first m terms of the numerator agree with the corresponding ones in the denominator. Hence $|H(j\Omega)|^2$ has $(2m-1)$ zero derivatives at $\Omega = 0$. Also the degrees of numerator and denominator of $H(\lambda)$ differ by $2(k-m)$ corresponding to the number of transmission zeros at $\Omega = \infty$ (i.e. $\omega = \omega_N/2$). A combination of k and m must be chosen to meet a given set of specifications.

Next, we note that we have only indicated the solutions to the approximation problem which satisfy certain optimality criteria, e.g. equiripple or maximal-flatness. This is a feature of our entire approach to the problem, not only in the present case but throughout this part of the book. However, in the design of FIR filters, there exist a number of popular techniques which are not in line with this approach, but are nevertheless characterized by their simplicity and can be useful.[15-17,82,83]

Finally, we note that due to the constrained form of an FIR transfer function, a high degree filter is needed to meet stringent specifications on the amplitude, by comparison with IIR filters. This is the price of inherent stability, simple realization, and the possibility of exact linear phase at all frequencies.

PROBLEMS

12.1 Design a low-pass Chebyshev digital filter with the following specifications:

Passband: 0–0.5 kHz, 0.1 dB ripple
Stopband edge: 0.7 kHz, attenuation ≥ 40 dB
Sampling frequency: 2 kHz

The realization is required in cascade form, and as a wave digital structure.

258

12.2 Consider the transfer function of a third order elliptic lumped filter

$$S_{21}(p) = \frac{0.314(p^2 + 2.806)}{(p + 0.767)(p^2 + 0.453p + 1.149)}$$

which gives a passband ripple of 0.5 dB, a minimum stopband attenuation of 21 dB, for $\omega_s/\omega_0 = 1.5$.

Use the given prototype function to design a digital filter with passband edge at 500 Hz and a sampling frequency of 3 kHz.

12.3 The following standard set of specifications are to be met by the transmit band-pass filter employed in a codec (coder–decoder) for PCM telephony,

Passband: 300–3200 Hz, 0.25 dB ripple
Stopband edge frequencies: 100 Hz and 4600 Hz with 32 dB minimum attenuation.

Design a Chebyshev digital filter which meets the above specifications. Use a lumped prototype function.

12.4 Design a low-pass Chebyshev wave digital filter as a cascade of wave UEs to meet the following specifications:

Passband: 0–3.4 kHz, 0.25 dB ripple
Stopband edge: 4.6 kHz, attenuation $\geq$30 dB
Sampling frequency: 32 kHz
50 Ω equal termination.

12.5 Consider the transfer function in (12.26) to (12.27) which possesses simultaneous maximally flat amplitude and group-delay at the origin ($\lambda = 0$). Show that the fifth degree function with $\alpha = 10$ meets the following set of specifications:

Passband: 0–2.25 kHz, attenuation $\leq$1 dB, delay variation $\leq$30 μs
Stopband: 3.6 kHz–5 kHz, attenuation $\geq$15 dB.

Realize the transfer function in cascade form.

Chapter 13

Approximation and Synthesis of a Class of Switched-capacitor Filters

13.1 INTRODUCTION

In this concluding chapter, a class of analogue sampled-data filters is introduced. It belongs to a category of networks which has come to be known as switched-capacitor filters. The basic building blocks are operational amplifiers (Op. Amps), analogue switches, and capacitors. The filter operates in the sampled-data mode, and its performance is determined by *ratios* of capacitor values not by absolute element values. These ratios can be realized very accurately and conveniently using modern silicon integrated circuit technology. In fact, the advent of this category of filters has dramatically raised the hopes of implementing, not only filters but entire information systems on a silicon chip.

In this chapter, a class of switched-capacitor filters is selected out of the large variety available to date. The choice of this particular class is dictated by a number of factors. In the first place, it is quite representative of the underlying design principles, and highlights the advantages of the entire category. In the second place, the selected class imitates the optimum sensitivity properties of a passive (commensurate distributed) filter. Finally it serves as an illustration of the power and depth of our central theme of passive network synthesis, thus adding another novel category of networks to the augmented theory without any need to introduce fundamentally new network-theoretic ideas. We shall see that the synthesis of this class of filters can be successfully accommodated into the general framework of the theoretical development in this book.

13.2 THE STATE-VARIABLE LADDER FILTER

Consider the general passive ladder shown in Fig. 13.1 where the branches are arbitrary impedances. Write the state equations of the ladder, relating the series currents to the shunt voltages. For the sake of specificity, we

259

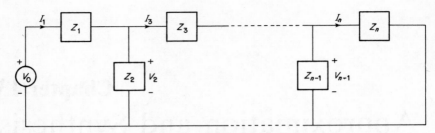

Fig. 13.1 General passive ladder.

assume that n is odd.

$$I_1 = Z_1^{-1}(V_0 - V_2)$$
$$V_2 = Z_2(I_1 - I_3)$$
$$I_3 = Z_3^{-1}(V_2 - V_4) \qquad (13.1)$$
$$\vdots \qquad \vdots$$
$$I_n = Z_n^{-1}V_{n-1}$$

The transfer function of interest is given by

$$H_{21} = \frac{I_n}{V_0} \qquad (13.2)$$

Regardless of the specific form of the branch impedances, any other circuit which implements the same set of equations (13.1) will produce the same transfer function. Note that we have deliberately refrained from using any specific frequency variable in (13.1).

Now, in seeking an active implementation of (13.1) all we need are some building blocks capable of providing the frequency dependence of the branches, as well as performing the mathematical operations of (13.1). A block diagram of the required implementation is shown in Fig. 13.2 and is known as the *state-variable leap-frog ladder* realization. This consists of differential-input boxes which possess voltage transfer functions T_1,

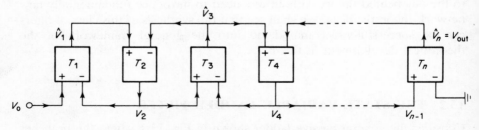

Fig. 13.2 State-variable (leap-frog) equivalent of the ladder of Fig. 13.1.

$T_2, \ldots, T_n$ connected such that

$$\hat{V}_1 = T_1(V_0 - V_2)$$
$$V_2 = T_2(\hat{V}_1 - \hat{V}_3)$$
$$\hat{V}_3 = T_3(V_2 - V_4) \qquad (13.3)$$
$$\vdots \quad \vdots$$
$$\hat{V}_n = T_n V_{n-1}$$

Let the currents $I_1, I_3, I_5, \ldots, I_n$ in (13.1) be simulated by the voltages $\hat{V}_1$, $\hat{V}_3, \hat{V}_5, \ldots, \hat{V}_n$ in (13.3). The internal structure of each box in Fig. 13.2 is chosen such that

$$T_1 = \alpha Z_1^{-1}$$
$$T_2 = \alpha^{-1} Z_2$$
$$T_3 = \alpha Z_3^{-1} \qquad (13.4)$$
$$\vdots \quad \vdots$$
$$T_n = \alpha Z_n^{-1}$$

where α is a constant. Then the transfer function of the leap-frog ladder of Fig. 13.2

$$\hat{H}_{21} = \frac{V_{\text{out}}}{V_0} \qquad (13.5)$$

only differs from H_{21} of the passive ladder by a constant. Consequently, given a ladder of the general form in Fig. 13.1, with the specific type and values of elements, it is possible to determine the state variable simulation provided we can find the necessary active building blocks with transfer functions satisfying (13.4). Conversely, if we decide on certain types of building blocks for the boxes in Fig. 13.2 we can find the equivalent ladder.

Now starting from Fig. 13.2, two basic types of building blocks are used as the boxes.[18,84,85] These are shown in Fig. 13.3 and are known as a lossless discrete integrator (LDI) and a damped discrete integrator (DDI). These are unfortunate misnomers which emerged in the early days of sampled-data filters and are still used extensively. So, for no other reason except to avoid confusing the reader, they are used here.

Type A This is the circuit of Fig. 13.3(a) and is the LDI. The Op.Amp. is assumed ideal and the switches are driven by a biphase clock with non-overlapping pulses of period T as shown in Fig. 13.4. The capacitor C_A is switched to the inputs V_{i1} and V_{i2} at $t = mT$ and to the Op.Amp. inputs at $t = (m + \frac{1}{2})T$. The output of the Op.Amp. is sampled at $t = (m + \frac{1}{2})T$. Thus

$$V_{\text{out}}(nT) = V_{\text{out}}\{(n-1)T\} + \frac{C_A}{C_B}(V_{i1}\{(n-\tfrac{1}{2})T\} - V_{i2}\{(n-\tfrac{1}{2})T\})$$

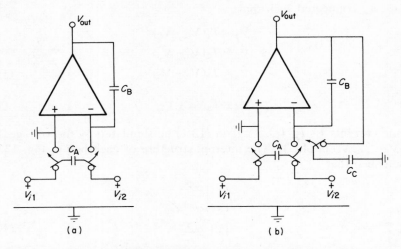

Fig. 13.3   The basic building blocks for a class of switched-capacitor filters. (a) Type A: lossless discrete integrator (LDI), (b) Type B: damped discrete integrator (DDI).

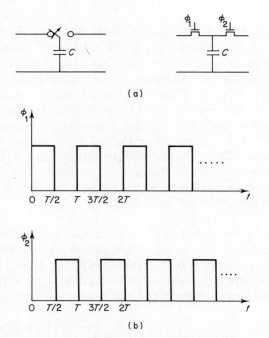

Fig. 13.4   (a) A switched-capacitor, (b) the biphase block driving the switches.

Taking the z-transform $(z = e^{Tp})$

$$V_{\text{out}}(z) = z^{-1} V_{\text{out}}(z) + z^{-1/2} \frac{C_A}{C_B} \{V_{i1}(z) - V_{i2}(z)\}$$

Therefore, this building block has the transfer function

$$
\begin{aligned}
T_A &= \frac{V_{\text{out}}(z)}{V_{i1}(z) - V_{i2}(z)} \\
&= \frac{z^{-1/2}}{\left(\dfrac{C_B}{C_A}\right)(1 - z^{-1})} \\
&= \frac{1}{2\left(\dfrac{C_B}{C_A}\right)\gamma}
\end{aligned}
\tag{13.6}
$$

where

$$\gamma = \sinh \frac{T}{2} p \tag{13.7}$$

Type B   This is the circuit of Fig. 13.3(b) which is the same as that of the LDI but with a feedback capacitor C_C. This is known as a damped discrete integrator (DDI). With the same sequence of switching as Type A we have

$$V_{\text{out}}(nT) = V_{\text{out}}\{(n-1)T\} + \frac{C_A}{C_B}(V_{i1}\{(n-1)T\} - V_{i2}\{(n - \tfrac{1}{2})T\})$$
$$- \frac{C_C}{C_B} V_{\text{out}}\{(n-1)T\}$$

and taking the z-transform

$$V_{\text{out}}(z) = z^{-1} V_{\text{out}}(z) + \frac{C_A}{C_B} z^{-1/2} \{V_{i1}(z) - V_{i2}(z)\}$$
$$- \frac{C_C}{C_B} z^{-1} V_{\text{out}}(z)$$

Therefore, this building block has the transfer function

$$
\begin{aligned}
T_B &= \frac{V_{\text{out}}(z)}{V_{i1}(z) - V_{i2}(z)} \\
&= \frac{z^{-1/2}}{\dfrac{C_B}{C_A}\left\{1 - z^{-1}\left(1 - \dfrac{C_C}{C_A}\right)\right\}} \\
&= \frac{1}{\left(\dfrac{2C_B - C_C}{C_A}\right)\gamma + \dfrac{C_C}{C_A}\mu}
\end{aligned}
\tag{13.8}
$$

where $\mu = \cosh (T/2)p$. \hfill (13.9)

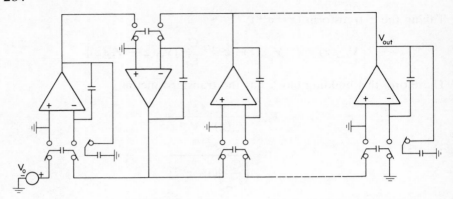

Fig. 13.5   The switched-capacitor state variable (leap-frog) ladder filter.

Now let these building blocks be used as the boxes in Fig. 13.2 such that the *first* and *last* boxes are *Type B* while all the *internal ones* are *Type A*. The resulting network takes the form shown in Fig. 13.5 and is clearly of the analogue sampled-data type. It is called a *switched-capacitor state-variable ladder*. Examination of (13.1) and (13.4) shows that the switched-capacitor network has the equivalent network of Fig. 13.6 where

$$L_k(\text{or } C_k) = 2\left(\frac{C_B}{C_A}\right)_k, \qquad k = 2 \rightarrow (n-1) \tag{13.10a}$$

$$L_{1,n} = \left(\frac{2C_B - C_C}{C_C}\right)_{1,n} \tag{13.10b}$$

$$R_{1,2} = \left(\frac{C_C}{C_A}\right)_{1,n} \tag{13.10c}$$

At first glance the equivalent network of Fig. 13.6 appears rather peculiar. Viewed as a doubly terminated ladder two-port, the elements have frequency dependence $L_k\gamma$, $1/C_k\gamma$ and the terminations are also frequency dependent of the form $R_1\mu$, $R_2\mu$. We shall see in a moment that the transfer function of this network is very familiar!

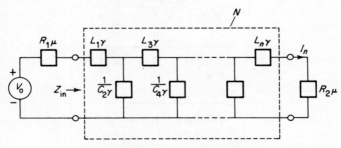

Fig. 13.6   Equivalent network.

Consider Fig. 13.6 and let the two-port N be described by a polynomial transmission matrix

$$[t(\gamma)] = \begin{bmatrix} a(\gamma) & b(\gamma) \\ c(\gamma) & d(\gamma) \end{bmatrix} \tag{13.11}$$

Viewed abstractedly as a polynomial transmission matrix of an odd degree ladder composed of elements of the indicated frequency dependence, we have

$$\begin{aligned} a(\gamma) &= a_{n-1}(\gamma) & \text{all even} \\ b(\gamma) &= b_n(\gamma) & \text{all odd} \\ c(\gamma) &= c_{n-2}(\gamma) & \text{all odd} \\ d(\gamma) &= d_{n-1}(\gamma) & \text{all even} \end{aligned} \tag{13.12}$$

Next, obtain the transfer function I_n/V_0 of the equivalent network.

$$H_{21} = \frac{I_n}{V_0} = \frac{1}{R_1 \mu a(\gamma) + b(\gamma) + R_1 R_2 \mu^2 c(\gamma) + R_1 \mu d(\gamma)} \tag{13.13}$$

and using

$$\gamma = \lambda / \sqrt{1 - \lambda^2} \tag{13.14}$$

$$\mu = 1 / \sqrt{1 - \lambda^2} \tag{13.15}$$

where

$$\lambda = \tanh \frac{T}{2} p \tag{13.16}$$

we have

$$H_{21}(\lambda) = \frac{(1 - \lambda^2)^{n/2}}{P_n(\lambda)} \tag{13.17}$$

which is the transfer function realized by the switched-capacitor network of Fig. 13.5 as a voltage transfer ratio V_{out}/V_0.

But the transfer function in (13.17) is identical in form to the transfer function of a cascade of UEs!, see Chapter 8, Section 8.5, expression (8.35). The solutions to the corresponding amplitude approximation problem were given in Section 11.2.3. Nevertheless the equivalent network of Fig. 13.6 is *not a resistively terminated lossless two-port*, so we require a modified synthesis technique. To this end, consider the process of impedance-scaling the equivalent network of Fig. 13.6 by the factor $1/\mu$. This results in the resistively terminated lossless two-port shown in Fig. 13.7 which is called the auxiliary network. This is taken as the starting-point in performing the synthesis of the original network of Fig. 13.6. The reason is that the standard synthesis technique (see Section 6.3.2) is directly applicable to the

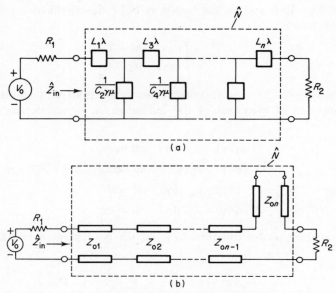

Fig. 13.7 Auxiliary network. (a) Obtained by impedance-scaling Fig. 13.6 by $1/\mu$, (b) alternative form as a cascade of $(n-1)$ U.E.s and a stub

auxiliary network. Its transfer function is given by

$$\hat{H}_{21} = \mu H_{21}$$
$$= \frac{(1-\lambda^2)^{(n-1)/2}}{P_n(\lambda)} \tag{13.18}$$

describing a cascade of $(n-1)$ UEs and one transmission zero at $\lambda = \infty$. If we denote the transducer power gain of the auxiliary network by $|S_{21}|^2$ then

$$|\hat{H}_{21}|^2 = |S_{21}|^2/4R_1R_2 \tag{13.19}$$

It also follows that if the input impedance $\hat{Z}_{in}$ of the auxiliary network is calculated using Darlington's procedure of Section 6.3.2, for any prescribed $|S_{21}|^2$, then the input impedance of the original network of Fig. 13.6 can be obtained from

$$Z_{in} = \mu \hat{Z}_{in} \tag{13.20}$$

Before discussing the approximation and synthesis we note that the necessary and sufficient condition for stability is that $P_n(\lambda)$ in (13.17) be strictly Hurwitz.

13.3 APPROXIMATION AND SYNTHESIS[84,85]

Comparing the transfer function (13.17) and that of (11.13) it follows that we already know the solutions to the amplitude approximation problem.

Thus, using (11.18) for a maximally flat response around the origin and one zero-derivative at $(T/2)\omega = \pi/2$,

$$|H_{21}|^2 = \frac{K}{1 + \left\{\dfrac{\sin \dfrac{T}{2} \omega}{\sin \dfrac{T}{2} \omega_0}\right\}^{2n}} \qquad (13.21)$$

where ω_0 is the 3-dB point. Clearly

$$\frac{T}{2} \omega = \pi \frac{\omega}{\omega_N}, \quad \frac{T}{2} \omega_0 = \pi \frac{\omega_0}{\omega_N} \qquad (13.22)$$

where ω_N is the radian sampling frequency. For an equiripple response up to ω_0 we have by (11.19),

$$|H_{21}|^2 = \frac{K}{1 + \varepsilon^2 T_n^2 \left\{\dfrac{\sin \dfrac{T}{2} \omega}{\sin \dfrac{T}{2} \omega_0}\right\}} \qquad (13.23)$$

Thus, for the auxiliary network in Fig. 13.7, assuming $R_1 = 1\,\Omega$, we use (13.18) and (13.19) to obtain

$$|S_{21}|^2 = \frac{4KR_2 \cos^2 \dfrac{T}{2} \omega}{1 + \left\{\dfrac{\sin \dfrac{T}{2} \omega}{\sin \dfrac{T}{2} \omega_0}\right\}^{2n}} \qquad (13.24)$$

for a maximally flat response of the original network and

$$|S_{21}|^2 = \frac{4KR_2 \cos^2 \dfrac{T}{2} \omega}{1 + \varepsilon^2 T_n^2 \left\{\dfrac{\sin \dfrac{T}{2} \omega}{\sin \dfrac{T}{2} \omega_0}\right\}} \qquad (13.25)$$

for an equiripple passband response of the original network. Now, in order to obtain the input impedance of the original network of Fig. 13.6 we must first proceed along the lines of Darlington's synthesis, in Section 6.3.2 for the auxiliary network of Fig. 13.7, since this technique applies to resistively terminated lossless two-ports. Having found the input impedance of this network (13.20) can be used to determine the input impedance of the original network of Fig. 13.6. Thus, from (13.24) and (13.25) we evaluate

$$|S_{11}|^2 = 1 - |S_{21}|^2 \qquad (13.26)$$

and the input reflection coefficient of the auxiliary network $S_{11}(\lambda)$ is obtained by factorization of (13.26) and selecting the left half λ-plane poles, i.e.

$$S_{11}(\lambda) = \frac{N(\lambda)}{D(\lambda)} \qquad (13.27)$$

where $D(\lambda)$ is a strictly Hurwitz polynomial in λ.

Once S_{11} is determined the input impedance of the original (realized) network is obtained using (13.27) and (13.20). Thus

$$Z_{in} = \mu \hat{Z}_{in} = \mu \frac{1 + S_{11}}{1 - S_{11}} \qquad (13.28)$$

Using

$$\mu^2 = (1 + \gamma^2), \qquad \lambda = \gamma/\mu \qquad (13.29)$$

Z_{in} may be put in the form

$$Z_{in} = \frac{\mu a_{n-1}(\gamma) + b_n(\gamma)}{\mu c_{n-2}(\gamma) + d_{n-1}(\gamma)} \qquad (13.30)$$

where a, b, c, and d are polynomials in γ which can be identified with those in (13.12). The synthesis technique then suggests itself as performing the continued fraction expansion of Z_{in} around γ, with μ treated as a constant. The coefficients of γ in the expansion give L_1, C_2, L_3, ... the element values in the network of Fig. 13.6. Subsequently the capacitance ratios can be obtained from (13.10).

Example 13.1   Consider the calculation of the element values for a seventh order low-pass Chebyshev filter with 0.05 dB passband ripple and passband edge at 3.4 kHz for a clock frequency of 28 kHz. In this case we have

$$f_0/f_N = \omega_0/\omega_N = 0.12$$

$$\sin(\pi f_0/f_N) = 0.37$$

Therefore from (13.23) and (13.24) with $\theta \equiv \pi\omega/\omega_N$, $K = 1/4$, and $R_2 = 1$,

$$|H_{21}|^2 = \frac{1/4}{1 + 0.01 T_7^2\left(\dfrac{\sin\theta}{0.37}\right)}$$

$$|S_{21}|^2 = \frac{\cos^2\theta}{1 + 0.01 T_7^2\left(\dfrac{\sin\theta}{0.37}\right)}$$

$$|S_{11}|^2 = \frac{\sin^2\theta + 0.01 T_7^2\left(\frac{\sin\theta}{0.37}\right)}{1 + 0.01 T_7^2\left(\frac{\sin\theta}{0.37}\right)}$$

The above expression is factored in the λ-domain so that $\hat{Z}_{in}$ is obtained from (13.28) as

$$\hat{Z}_{in}(\lambda) = \frac{\begin{aligned}&1 + 16.55\lambda + 112.8\lambda^2 + 573.04\lambda^3 + 1660.94\lambda^4 + 4716.78\lambda^5 \\ &\qquad\qquad + 5854.26\lambda^6 + 10391.77\lambda^7\end{aligned}}{\begin{aligned}&1 + 12.27\lambda + 84.28\lambda^2 + 301.87\lambda^3 + 948.69\lambda^4 + 1366.66\lambda^5 \\ &\qquad\qquad + 2425.928\lambda^6\end{aligned}}$$

Then using (13.28) $Z_{in}(\lambda)$ is put in the form given by (13.30)

$$Z_{in}(\gamma, \mu) = \frac{\begin{aligned}&15704.13\gamma^7 + 7629.00\gamma^6\mu + 5924.49\gamma^5 + 1889.54\gamma^4\mu \\ &\qquad + 622.8\gamma^3 + 115.80\gamma^2\mu + 16.54\gamma + \mu\end{aligned}}{\begin{aligned}&3459.9\gamma^6 + 1680.8\gamma^5\mu + 1120.25\gamma^4 + 326.41\gamma^3\mu \\ &\qquad + 87.27\gamma^2 + 12.27\gamma\mu + 1.0\end{aligned}}$$

Performing the continued fraction expansion we obtain for the element values of Fig. 13.6

$$L_1 = 4.53, \quad C_2 = 4.12, \quad L_3 = 5.18, \quad C_4 = 4.4$$
$$L_5 = 4.77, \quad C_6 = 3.74, \quad L_7 = 2.05, \quad R_2 = 1$$

The actual capacitor ratios are then obtained from the above values and (13.10).

13.4 PRACTICAL CONSIDERATIONS

The real merit of these switched capacitor ladder filters, is twofold. First, the performance of the filter is determined by capacitance ratios and not by absolute element values. These ratios can be very accurately and conveniently realized using present-day silicon integrated circuit technologies; the absolute capacitor values can be reduced to increase the density of elements on a single chip. The other advantage is that the leap-frog ladder filter is modelled on a passive (distributed) prototype, and like wave digital filters, retains the good sensitivity property of passive filters.

When designing switched-capacitor filters using silicon integrated circuit technology, the effect of stray parasitic capacitances must be taken into account. Attention has, therefore, been given to the use of building blocks which perform the same function as those of Fig. 13.3 but are *insensitive* to parasitic capacitances. This objective can be easily achieved in the following way.

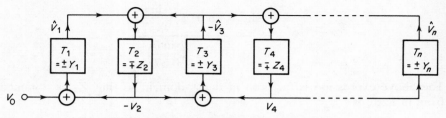

Fig. 13.8 Modified simulation of the same equations (13.1).

(a) Modify the state-variable leap-frog block diagram of Fig. 13.2 as shown in Fig. 13.8.
(b) Instead of the building blocks of Fig. 13.3, use the modified ones of Figs 13.9 and 13.10.

Modified Type A These are the circuits of Fig. 13.9, with transfer functions

$$T_A = \frac{z^{-1}}{\frac{C_B}{C_A}(1-z^{-1})} = \frac{z^{-1/2}}{2\frac{C_B}{C_A}\gamma} \tag{13.31}$$

for the circuit of Fig. 13.9(a), and

$$\hat{T}_A = \frac{-1}{\frac{C_B}{C_A}(1-z^{-1})} = \frac{-z^{1/2}}{2\frac{C_B}{C_A}\gamma} \tag{13.32}$$

for the network of Fig. 13.9(b).

Modified Type B These are the circuits of Fig. 13.10 with transfer functions

$$T_B = \frac{z^{-1}}{\frac{C_B}{C_A}\left\{1-z^{-1}+\frac{C_C}{C_B}\right\}}$$

$$= \frac{z^{-1/2}}{\left(\frac{2C_B+C_C}{C_A}\right)\gamma + \frac{C_C}{C_A}\mu} \tag{13.33}$$

for the circuit of Fig. 13.10(a); and

$$\hat{T}_B = \frac{-1}{\frac{C_B}{C_A}\left\{1-z^{-1}+\frac{C_C}{C_B}\right\}}$$

$$= \frac{-z^{1/2}}{\left(\frac{2C_B+C_C}{C_A}\right)\gamma + \frac{C_C}{C_A}\mu} \tag{13.34}$$

for the circuit of Fig. 13.10(b).

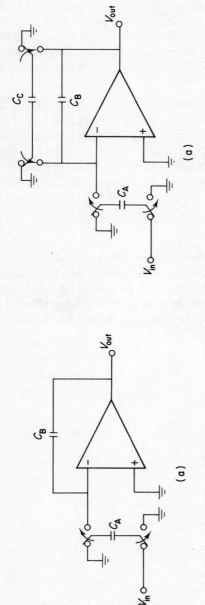

Fig. 13.9 Strays-insensitive modified Type A building block.
(a) positive, (b) negative.

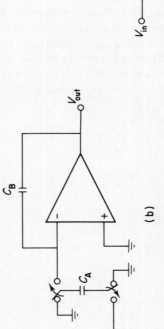

Fig. 13.10 Strays-insensitive modified Type B building
block. (a) positive, (b) negative.

(c) In the realization of Fig. 13.8, using the modified building blocks, two points must be observed. First, positive and negative building blocks alternate. Secondly, the *first* and *last* building blocks are of *Type B* while the rest are of *Type A*. Clearly the summing operation is easily realized by adding another capacitor of the same value C_A to every building block, i.e. between the voltage to be summed and the node where the original C_A is connected.

Having modified the procedure as outlined above, it is a simple matter to show that the resulting filter has the same equivalent network of Fig. 13.6 except that all impedances are scaled by $z^{1/2}$. This multiplies the transfer function in (13.17) by a factor $z^{-1/2}$, which of course has no effect on the amplitude response of the filter. The resulting structure, however, has the advantage of being completely insensitive to stray parasitic capacitances inherent in the manufacture process.

13.5 CONCLUSION

A representative class of switched-capacitor ladder filters has been discussed. The synthesis technique for this class has been given, and the chosen filter imitates the low sensitivity properties of a passive (commensurate distributed) counterpart. Only low-pass filters have been considered but the reader may consult the references for other types, as well as different design techniques of switched-capacitor filters.[86-94]

Appendices

A.1 HERMITIAN AND QUADRATIC FORMS

Consider a square $n \times n$ matrix $[A(p)]$ such that $[\tilde{A}(p)] = [A(p)]$. Then $[A]$ is called *Hermitian*. Let $[x(p)]$ be a column matrix

$$[x(p)] = \begin{bmatrix} x_1(p) \\ x_2(p) \\ \vdots \\ x_n(p) \end{bmatrix} \tag{A.1}$$

The expression

$$[\tilde{x}][A][x] = f \tag{A.2}$$

is called a *Hermitian form*. If $[A]$ is real constant, then $[\tilde{A}] = [A]' = [A]$, i.e. $[A]$ is real symmetric, and the expression in (A.2) is called a *symmetric quadratic form*. f is called a quadratic form even if $[A]$ is not symmetric.

If f is *strictly positive* (>0) for arbitrary values of $x_1, x_2, \ldots, x_n$, not all zero, the Hermitian (or quadratic) form is called *positive definite*. If f is *non-negative* (≥ 0) for arbitrary values of $x_1, x_2, \ldots, x_n$, not all zero, the Hermitian (or quadratic) form is called *positive semi-definite*.

The *matrix* $[A]$ of a Hermitian (or quadratic) form is called *positive definite* or *semi-definite* according to whether the form is positive definite or semi-definite, respectively. Write a real symmetric matrix as

$$[A] = \begin{bmatrix} a_{11} & a_{12} & \cdots & a_{1n} \\ a_{12} & a_{22} & & \vdots \\ \vdots & & & \\ a_{1n} & & \cdots & a_{nn} \end{bmatrix} \tag{A.3}$$

The necessary and sufficient conditions for $[A]$ to be positive definite are that all the *leading* principal minors of $[A]$ be strictly positive (>0). For $[A]$ to be positive semi-definite, it is necessary and sufficient that *all* the principal minors (not only the leading ones) be non-negative (≥ 0). Similar conditions hold for a Hermitian matrix.

A.2 ANALYTIC FUNCTIONS

A function $f(p)$ which is one-valued and differentiable at every point of a domain is said to be *analytic* (or regular, or holomorphic) in the domain.

273

274

$f(p)$ may be differentiable in a domain, save possibly for a finite number of points. These points are called *singularities* of $f(p)$. The Weierstrassan development of complex variable theory begins by *defining* an analytic function as one which is expansible in power series.

A polynomial $Q(p)$ may be regarded as a power series which converges for all p. Therefore, a rational function

$$f(p) = \frac{a_0 + a_1 p + \ldots a_m p^m}{b_0 + b_1 p + \ldots b_n p^n} \tag{A.4}$$

is analytic at all points of the p-plane at which the denominator does not vanish, with a possible pole at ∞. Thus, a rational function is one whose only singularities are poles.

The maximum modulus theorem states that: if a function $f(p)$ is analytic within and on the boundary of a domain, then the modulus of $f(p)$ has its maximum value on the boundary and not at any interior point.

The minimum real-part theorem states that: the real part of an analytic function $f(p)$ attains its minimum value on the boundary of the domain of analyticity, not at any interior point.

A.3 RELATIONSHIP BETWEEN A POSITIVE REAL AND A BOUNDED REAL FUNCTION

Let $Z(p)$ be a p.r.f. and write

$$S(p) = \frac{Z(p) - 1}{Z(p) + 1} \tag{A.5}$$

The mapping between the Z-plane and S-plane is shown in Fig. A.1. From (A.5) it is clear that S is real for p real since Z is assumed p.r. The right half of the Z-plane is mapped into the inside of the unit circle in the S-plane and the jx-axis becomes the contour of the unit circle. Since Z is p.r., right half-plane values of p map on to right half-plane values of Z, and so are mapped inside the unit circle of the S-plane. For $p = j\omega$, $Z(j\omega)$ takes on values in the right half Z-plane or on the jx-axis, and these fall inside or on

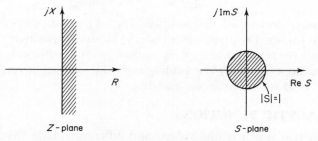

Fig. A.1 The mapping between the Z and S-planes.

the unit circle in the S-plane. Hence $|S(j\omega)| \leq 1$. Next consider the poles of S. These occur where $Z = -1$. But values of Z where this is the case, cannot lie in the closed right half-plane since Z is p.r. Therefore S must be analytic in the closed right half-plane, i.e. including the $j\omega$-axis, in particular the point at infinity. Therefore S satisfies Theorem 2.2 and is bounded real.

To prove the converse, let S be a bounded real function and write

$$Z = \frac{1+S}{1-S}$$

$$= \frac{1+S}{1-S} \cdot \frac{1-S_*}{1-S_*} \tag{A.6}$$

then

$$\operatorname{Re} Z(j\omega) = \frac{1 - |S(j\omega)|^2}{|1 - S(j\omega)|^2} \tag{A.7}$$

But S is bounded real and satisfies $|S(j\omega)|^2 \leq 1$. Thus $\operatorname{Re} Z(j\omega) \geq 0$. Since $S(p)$ is analytic in the closed right half-plane, including infinity, then according to the maximum modulus theorem $|S(p)|$ has its maximum on the $j\omega$-axis so that $\operatorname{Re} Z(p) > 0$ in $\operatorname{Re} p > 0$. Thus $Z(p)$ is p.r.

A.4 PROOF OF RICHARDS' THEOREM

Consider the function

$$\hat{Z}_1(\lambda) = \frac{Z_1(\lambda)}{Z(1)} = \frac{Z(\lambda) - \lambda Z(1)}{Z(1) - \lambda Z(\lambda)} \tag{A.8}$$

so that $\hat{Z}_1(\lambda)$, hence $Z_1(\lambda)$ is p.r. if and only if

$$\hat{S}_1(\lambda) = \frac{\hat{Z}_1(\lambda) - 1}{\hat{Z}_1(\lambda, \ 1)} \tag{A.9}$$

is bounded real. Calculate

$$\hat{S}_1(\lambda) = \left\{ \frac{1+\lambda}{1-\lambda} \right\} \left\{ \frac{Z(\lambda)/Z(1) - 1}{Z(\lambda)/Z(1) + 1} \right\}$$

$$= \left\{ \frac{1+\lambda}{1-\lambda} \right\} \Gamma(\lambda) \tag{A.10}$$

But since $Z(\lambda)$ is p.r., so is $Z(\lambda)/Z(1)$, because $Z(1)$ is guaranteed positive. Therefore Γ is bounded real. Moreover the numerator of Γ vanishes at $\lambda = 1$, therefore, it contains the factor $(1 - \lambda)$. Similarly the denominator of Γ contains the factor $(1 + \lambda)$. It follows that the function which multiplies Γ in (A.10) cancels out, and since Z is p.r., Γ, hence $\hat{S}_1$ is bounded real. Consequently $\hat{Z}_1$, and therefore Z_1 is p.r. of degree at most equal to that of

Z. In (A.8) the numerator and denominator of Z_1 contain a common factor $(1-\lambda)$ since they both vanish at $\lambda = 1$. If, in addition $Z(\lambda)$ satisfies (8.31) then $Z(1) = -Z(-1)$ and the numerator and denominator of Z_1 have the additional common factor $(1+\lambda)$. Therefore if Z is a p.r.f. satisfying (8.31) we may extract a UE leaving a remainder Z_1 that is p.r., and of lower degree by unity than Z, after the cancellation of a common factor $(1-\lambda^2)$ in (A.8).

References

1. E. G. Phillips, *Functions of a Complex Variable*, Oliver and Boyd, 1972.
2. F. R. Gantmacher, *The Theory of Matrices*, Vol. 1, Chelsea Publ. Co. NY, 1960.
3. E. Guillemin, *The Mathematics of Circuit Analysis*, MIT Press, Cambridge, Mass. USA, 9th Ed., 1969.
4. E. Guillemin, *Introductory Circuit Theory*, J. Wiley and Sons. NY. 1956.
5. J. O. Scanlan and R. Levy, *Circuit Theory*, Vol. 1, Oliver and Boyd, 1970.
6. F. F. Kuo, *Network Analysis and Synthesis*, J. Wiley and Sons, NY, 1966.
7. M. E. Van Valkenburg, *Network Analysis*, Prentice-Hall, Englewood Cliffs, NJ, 1972.
8. J. D. Rhodes, *Theory of Electrical Filters*, J. Wiley and Sons, 1976.
9. G. Daryanani, *Principles of Active Network Synthesis and Design*, J. Wiley and Sons, NY, 1976.
10. L. P. Huelsman and P. E. Allen, *Introduction to the Theory and Design of Active Filters*, McGraw-Hill, 1980.
11. M. Ghausi and K. Laker, *Modern Filter Design: Active RC and Switched-Capacitor*, Prentice-Hall, Englewood Cliffs, NJ, 1981.
12. G. S. Moschytz and P. Horn, *Active Filter Design Handbook*, J. Wiley and Sons, 1981.
13. G. C. Temes and S. K. Mitra (Eds), *Modern Filter Theory and Design*, J. Wiley and Sons, NY, 1973.
14. H. H. Ernyei, 'Mechanical filter design with lumped element prototypes'. *IEEE Trans. Circuits and Systems*, **CAS-30,** No. 2, pp. 89–107, Feb. 1983.
15. H. W. Schussler, *Digitale Systeme zur Signalverarbeitung*, Springer-Verlag, Berlin, 1973.
16. L. Rabiner and B. Gold, *Theory and Application of Digital Signal Processing*, Prentice-Hall Inc., Englewood Cliffs, NJ, 1975.
17. A. Antoniou, *Digital Filters, Analysis and Design*, McGraw-Hill, NY, 1979.
18. G. M. Jacobs *et al.*, Design techniques for MOS switched-capacitor ladder Filters, *IEEE Trans. Circuits and Systems*, **CAS-25,** No. 12, pp. 1014-1021, Dec. 1978.
19. H. J. Carlin and A. B. Giordano, *Network Theory*, Prentice-Hall Inc., Englewood Cliffs, NJ, 1964.
20. V. Belevitch, *Classical Network Theory*, Holden-Day Inc., San Francisco, 1968.
21. N. Balabanian, T. Bickart, and S. Seshu, *Electrical Network Theory*, J. Wiley and Sons, NY, 1969.
22. B. D. H. Tellegen, 'The gyrator, a new electric network element', *Philips Res. Rep.* 3, pp. 81–101, April 1948.
23. B. D. H. Tellegen, 'A general network theorem with applications', *Philips Res. Rep.* 7, pp. 259–269, Aug. 1952.
24. E. Guillemin, *Synthesis of Passive Networks*, J. Wiley and Sons, NY, 1965.
25. M. E. Van Valkenburg, *Introduction to Modern Network Synthesis*, J. Wiley and Sons, NY, 1964.

278

26. E. I. Jury, *Inners and Stability of Dynamic Systems*, J. Wiley and Sons, NY, 1974.
27. R. M. Foster, 'A reactance theorem', *Bell System Tech. J.*, **3**, pp. 259–267, April 1924.
28. D. C. Youla, 'A tutorial exposition of some key network—theoretic ideas underlying classical insertion—loss filter design', *Proc. IEEE*, **59**, No. 5, pp. 760–799, May 1971.
29. B. Brune, 'Synthesis of a finite two-terminal network whose driving-point impedance is a prescribed function of frequency', *J. Math. Phys.*, **10**, No. 3, pp. 191–236, 1931.
30. S. Darlington, 'Synthesis of reactance 4-poles which produce prescribed insertion loss characteristics', *J. Math. Phys.*, **18**, No. 4, pp. 257–353, 1939.
31. D. C. Youla, 'A new theory of cascade synthesis', *IRE Trans., Circuit Theory*, **CT-8**, pp. 244–260, Sept. 1966.
32. J. O. Scanlan and J. D. Rhodes, 'Unified theory of dascade synthesis', *Proc. IEE*, **17**, No. 4, pp. 665–670, April 1970.
33. R. Bott and R. J. Duffin, 'Impedance synthesis without the use of transformers', *J. Appl. Phys.*, **20**, No. 8, p. 816, 1949.
34. A. D. Fialkow, 'Inductance, capacitance networks terminated in resistance', *IEEE Trans. Circuits and Systems*, **CAS-26**, No. 8, pp. 603–641, Aug. 1979.
35. A. Fialkow, 'A note on LC-R ladder synthesis of filters', *IEEE Trans. Circuits and Systems*, **CAS-29**, No. 5, pp. 331–333, May 1982.
36. T. Fujisawa, 'Realisability theorem for mid-series or mid-shunt low-pass ladders without mutual induction, *IRE Trans. Circuit Theory*, **CT-2**, pp. 320–325, Dec. 1955.
37. H. J. Carlin, 'Distributed circuit design with transmission-line elements', *Proc. IEEE*, **59**, No. 7, pp. 1059–1081, July 1971.
38. J. O. Scanlan, 'Theory of microwave coupled-line networks', *Proc. IEEE*, **68**, No. 2, pp. 209–231, Feb. 1980.
39. P. I. Richards, 'Resistor-transmission-line networks', *Proc. IRE*, **30**, pp. 217–220, Feb. 1948.
40. J. O. Scanlan and J. D. Rhodes, 'Cascade synthesis of distributed networks', *Proc. 1966 Symposium on Generalised Networks, Brooklyn, N.Y.*, Polytechnic Inst. of Brooklyn Press, pp. 227–255.
41. P. I. Richards, 'General impedance-function theory', *Quart. Appl. Math.*, **6**, pp. 21–29, 1948.
42. J. D. Rhodes, P. C. Marston, and D. C. Youla, 'Explicit solution for the synthesis of two-variable transmission line networks', *IEEE Trans. Circuit Theory*, **CT-20**, pp. 504–511, Sept. 1973.
43. R. Levy, 'A general equivalent circuit transformation for distributed networks', *IEEE Trans. Circuit Theory* **CT-12**, pp. 457–458, Sept. 1965.
44. G. L. Mathei, L. Young, and E. M. Jones, *Microwave Filters, Impedance-matching Networks and Coupling Structures*, McGraw-Hill, NY, 1964.
45. H. Uchida, *Fundamentals of Coupled-lines and Multiwire Antennas*, Tohoku University Electronics Series, Vol. 1, Sasaki Printing and Publ. Co. Ltd., Sendai, Japan, 1967.
46. A. Matsumoto (Ed.), *Microwave Filters and Circuits*, Academic Press, NY, 1970.
47. R. J. Wenzel, 'Exact theory of interdigital band-pass filters and related coupled structures', *IEEE Trans. Microwave Theory and Techniques*, **MTT-13**, pp. 559–575, Sept. 1965.
48. H. J. Carlin and W. Kohler, 'Direct synthesis of band-pass transmission line structures', *IEEE Trans. Microwave Theory and Techniques*, **MTT-13**, pp. 559–575, Sept. 1965.

49. J. D. Rhodes, 'Theory of generalised interdigital networks', *IEEE Trans. Circuit Theory*, **CT-16,** No. 2, pp. 280–288, Aug. 1969.
50. D. I. Porat and A. Barna, *Introduction to Digital Techniques* J. Wiley and Sons, NY, 1979.
51. E. I. Jury, *Theory and Application of the Z-Transform Method*, J. Wiley and Sons, NY, 1964.
52. H. J. Orchard, 'Inductorless filters', *Electron. Lett.*, **2,** pp. 224–225, June 1966.
53. G. C. Temes and H. J. Orchard, 'First-order sensitivity and Worst Case analysis of doubly terminated reactance two-ports', *IEEE Trans. Circuit Theory*, **CT-20,** pp. 567–571, Sept. 1973.
54. H. J. Orchard, 'Loss sensitivity in singly and doubly terminated filters', *IEEE Trans. Circuits and Systems*, **CAS-26,** No. 5, pp. 293–297, May 1979.
55. A. Fettweis, 'Digital filter structures related to classical filter networks', *Arch. Elektr. Ubertragung*, **25,** pp. 78–89, 1971.
56. A. Fettweis, 'Some principles of designing digital filters imitating classical filter structures', *IEEE Trans. Circuit Theory*, **CT-18,** pp. 314–316, March 1971.
57. A. Sedlmeyer and A Fettweis, 'Digital filters with true ladder configuration', *International Journal of Circuit Theory and Applications*, **1,** No. 1, pp. 5–10, March 1973.
58. M. Fahmy, 'Digital realisation of C and D-type sections', *International Journal of Circuit Theory and Applications*, **3,** pp. 395–402, 1975.
59. R. Nouta, 'Wave digital cascade synthesis', *International Journal of Circuit Theory and Applications*, **3,** No. 3, pp. 231–248, Sept. 1975.
60. S. S. Lawson, 'On a generalisation of the wave digital filter concept', *International Journal of Circuit Theory and Applications*, **6,** No. 2, pp. 107–120, April 1978.
61. R. W. Daniels, *Approximation Methods for Electronic Filter Design*, McGraw-Hill, NY, 1974.
62. M. Abramowitz and I. A. Stegun (Eds), *Handbook of Mathematical Functions*, Dover Publ. Inc., NY, 1970.
63. P. Amstutz, 'Elliptic approximation and elliptic filter design on small computers', *IEEE Trans. Circuits and Systems*, **CAS-25,** No. 12, pp. 1001–1011. Dec. 1978.
64. R. Saal, *Handbook of Filter design*, AEG Telefunken, 1979.
65. W. E. Thompson, 'Delay networks having maximally-flat frequency characteristics', *Proc. IEE*, **96,** pp. 487–490, Nov. 1949.
66. T. Henk, 'The generation of arbitrary phase polynomials by recurrence formulae', *International Journal of Circuit Theory and Applications*, **9,** No. 4, pp. 461–478, Oct. 1981.
67. P. H. Halpern, 'Solution of flat time delay at finite frequencies', *IEEE Trans. Circuit Theory*, **CT-18,** pp. 241–246, March 1971.
68. K. K. Pang and P. A. Kirton, 'Optimum delay filter characteristics', *International Journal of Circuit Theory and Applications*, **10,** No. 4, pp. 361–375, Oct. 1982.
69. H. J. Carlin and J. L. Wu, 'Amplitude selectivity versus constant delay in minimum-phase lossless filters', *IEEE Trans. Circuits and Systems*, **CAS-23,** No. 7, pp. 447–455, 1976.
70. S. O. Scanlan and H. Baher, 'Filters with maximally-flat amplitude and controlled delay responses', *IEEE Trans. Circuits and Systems*, **CAS-23,** No. 5, pp. 270–278, May 1976.
71. J. D. Rhodes and I. H. Zabalawi, 'Design of selective linear-phase filters with equiripple amplitude characteristics', *IEEE Trans. Circuits and Systems*, **CAS-25,** No. 12, pp. 989-1000, Dec. 1978.
72. J. D. Rhodes and I. H. Zabalawi, 'Selective linear-phase filters possessing a pair

280

of j-axis transmission zeros', *International Journal of Circuit Theory and Applications*, **10**, No. 3, pp. 251–263, July 1982.

73. P. Jarry, 'Transfer functions interpolating an ideal amplitude with arbitrary phase and delay', *International Journal of Circuit Theory and Applications*, **11**, No. 2, pp. 131–140, April 1983.

74. R. Levy, 'Tables of element values for the distributed low-pass prototype filter', *IEEE Trans. Microwave Theory and Techniques*, **MTT-13**, pp. 514–536, Sept. 1965.

75. T. A. Abele, 'Transmission line filters approximating a constant delay in a maximally-flat sense', *IEEE Trans. Circuit Theory*, **CT-14**, pp. 295–306, Sept. 1967.

76. H. J. Carlin and J. L. Wu, 'Linear phase commensurate line networks with amplitude selectivity, *Proc. Int. Symposium on Circuits and Systems, Munich*, pp. 536–539, 1976.

77. M. F. Fahmy, 'The use of Pade approximants in the derivation of distributed low-pass filters with simultaneous flat amplitude and delay characteristics', *International Journal of Circuit Theory and Applications*, **8**, No. 3, pp. 192–204, July 1980.

78. P. Jarry, 'Synthesis of some non-reciprocal filters employing the Jacobi polynomials', *International Journal of Circuit Theory and Applications*, **8**, No. 3, pp. 229–236, July 1980.

79. J. H. McClellan, T. W. Parks, and L. R. Rabinar, 'A computer program for designing optimum-FIR linear-phase digital filters', *IEEE Trans. Audio Electroacoust.*, **AU-21**, pp. 506–526, Dec. 1973.

80. A. Antoniou, 'Accelerated procedure for the design of equiripple non-recursive digital filters', *Proc. IEE*, **129**, Pt. G, No. 1, pp. 1–10, Feb. 1982.

81. O. Hermann, 'On the approximation problem in non-recursive digital filter design', *IEEE Trans. Circuit Theory*, **CT-18**, pp. 411–413, May 1971.

82. S. O. Scanlan and H. Baher, 'Analytic design for FIR filters', *Proceedings European Conference Circuit Theory and Design, Lausanne*, pp. 305–309, 1978.

83. J. W. Adams and A. N. Wilson Jr., 'A new approach to FIR digital filters with fewer multipliers and reduced sensitivity', *IEEE Trans. Circuits and Systems*, **CAS-30**, No. 5, pp. 277–283, May 1983.

84. S. O. Scanlan, 'Analysis and synthesis of switched-capacitor state-variable filters', *IEEE Trans. Circuits and Systems*, **CAS-28**, No. 2, pp. 85–92, Feb. 1981.

85. H. Baher and S. O. Scanlan, 'Stability and exact synthesis of low-pass switched-capacitor filters', *IEEE Trans. Circuits and Systems*, **CAS-29**, No. 7, pp. 488–492, July 1982.

86. A. Fettweis, 'Basic principles of switched-capacitor filters using voltage inverter switches', *Arch. Elektr. Ubertrag.*, **33**, pp. 13–19, Jan. 1979.

87. A. Fettweis, 'Switched-capacitor filters using voltage inverted switchers: further design principles', *Arch. Elektr. Ubertrag.*, **33**, pp. 107–114, Mar. 1979.

88. A. Fettweis *et al.*, 'MOS switched-capacitor filters using voltage inverter switches', *IEEE Trans. Circuits and Systems*, **CAS-27**, No. 6, pp. 527–538, June 1980.

89. K. Martin and A. S. Sedra, 'Exact design of switched-capacitor bandpass filters using coupled-biquad structures', *IEEE Trans. Circuits and Systems*, **CAS-27**, No. 6, pp. 469–475, June 1980.

90. M. S. Lee and C. Chang, 'Switched-capacitor filters using the LDI and bilinear transformations', *IEEE Trans. Circuits and Systems*, **CAS-28**, No. 4, pp. 265–270, April 1981.

91. M. S. Lee, G. C. Temes, C. Chang, and M. B. Ghaderi, 'Bilinear switched-capacitor ladder filters', *IEEE Trans. Circuits and Systems*, **CAS-28**, No. 8, pp. 811–822, Aug. 1981.

92. U. W. Brugger, D. C. Von Grunigen, and G. S. Moschytz, 'A comprehensive procedure for the design of cascaded switched-capacitor filters', *IEEE Trans. Circuits and Systems*, **CAS-28,** No. 8, pp. 803–810, Aug. 1981.

93. M. S. Lee and C. Chang, 'Low-sensitivity switched-capacitor ladder filters', *IEEE Trans. Circuits and Systems*, **CAS-27,** No. 6, pp. 475–480, June 1980.

94. M. B. Ghaderi, J. A. Nossek, and G. C. Temes, 'Narrow-band switched-capacitor bandpass filters', *IEEE Trans. Circuits and Systems*, **CAS-29,** No. 8, pp. 557–572, Aug. 1982.

95. H. Baher and S. O. Scanlan, 'Exact synthesis of bandpass switched-capacitor LDI ladder filters', *IEEE Trans. Circuits and Systems*, **CAS-31,** No. 4, April 1984.

Index